Advances in Anatomy, Embryology and Cell Biology publishes critical reviews and state-of-the-art surveys on all aspects of anatomy and of developmental, cellular and molecular biology, with a special emphasis on biomedical and translational topics.

The series publishes volumes in two different formats:

- Contributed volumes, each collecting 5 to 15 focused reviews written by leading experts
- Single-authored or multi-authored monographs, providing a comprehensive overview of their topic of research

217
Advances in Anatomy, Embryology and Cell Biology

More information about this series at
http://www.springer.com/series/102

Udo Rüb • Jean Paul G. Vonsattel
Helmut Heinsen • Horst-Werner Korf

The Neuropathology of Huntington's Disease: Classical Findings, Recent Developments and Correlation to Functional Neuroanatomy

Springer

Udo Rüb
Dr. Senckenbergisches
Chronomedizinisches Institut
Goethe University Frankfurt
Frankfurt
Germany

Jean Paul G. Vonsattel
Medical Center Neurological Institute
Columbia University
New York, NY
USA

Helmut Heinsen
Division Psychiatic Clinic Morphological
Brain Research Unit
Julius Maximilians University Würzburg
Würzburg
Germany

Department of Pathology
University of Sao Paulo Medical School
Sao Paulo
Brazil

Horst-Werner Korf
Dr. Senckenbergisches
Chronomedizinisches Institut
Goethe-University
Frankfurt
Hessen
Germany

ISSN 0301-5556 ISSN 2192-7065 (electronic)
Advances in Anatomy, Embryology and Cell Biology
ISBN 978-3-319-19284-0 ISBN 978-3-319-19285-7 (eBook)
DOI 10.1007/978-3-319-19285-7

Library of Congress Control Number: 2015949474

Springer Cham Heidelberg New York Dordrecht London

Printed on acid-free paper

Springer International Publishing AG Switzerland is part of Springer Science+Business Media (www.springer.com)

Abstract

Huntington's disease (HD) is a severe, autosomal dominantly inherited, gradually worsening neurological disorder, the clinical features of which were first described in 1863 by Irving W. Lyon and with additional details, in 1872, by George Huntington. Progress in molecular biological research has shown that HD is caused by meiotically unstable CAG-repeats in the mutated HD gene (the so-called IT 15 gene) on chromosome 4p16.3, which encodes the mutated protein huntingtin (Htt). This monograph provides a survey of the stepwise progress in neuropathological HD research made during a time period of more than hundred years, the currently known neuropathological hallmarks of HD, as well as their pathogenic and clinical relevance. Starting with the initial descriptions of the progressive degeneration of the neostriatum (i.e., caudate nucleus and putamen) as one of the key events in HD, the worldwide practiced Vonsattel HD grading system of striatal neurodegeneration will be outlined. Correlating qualitative and quantitative neuropathological data with characteristics pertaining to the functional neuroanatomy of the human brain, subsequent chapters will highlight the latest neuropathological HD findings: the area- and layer-specific neuronal loss in the cerebral neo- and allocortex, the neurodegeneration of select thalamic nuclei, the affection of the cerebellar cortex and the deep cerebellar nuclei, the involvement of distinct brainstem nuclei, and the pathophysiological relevance of these pathologies for the clinical phenotype of HD. Finally, the potential pathophysiological role of axonal transport deficits and impairments of the neuronal protein quality control machinery, the possible pathogenetic relevance of the various types of neuronal Htt aggregations, and morphologically altered nerve cells for the pathological process of HD are discussed. All findings provide clear evidence that HD represents a multisystem degenerative disease of the human brain affecting the striatum, cerebral neo- and allocortex, thalamus, brainstem, and cerebellum. The findings suggest that the disease process spreads via distinct neuronal pathways in an anterograde and/or retrograde fashion and that the brain neuropathology of the polyglutamine disease HD is more closely related to the polyglutamine ataxic disorders (spinocerebellar ataxias) than previously thought.

Acknowledgments

This project was supported by grants from the Dr. Senckenbergische Stiftung (Frankfurt/Main, Germany), the Deutsche Huntington Hilfe e.V., and the Huntington-Selbsthilfe Nordrhein-Westfalen e.V. The skillful assistance of D. von Meltzer (secretary), M. Babl (tissue processing and staining), M. Bouzrou (tissue preparation and staining), I. Szasz (graphics), and K. Seidel (graphics) is thankfully acknowledged.

Contents

Non-neuroanatomical Abbreviations

AA	Amino acid
AD	Alzheimer's disease
AGD	Argyrophilic grains disease
ALP	Autophagy-lysosomal pathway
ATP	Adenosine triphosphate
BAEP	Brainstem auditory evoked potentials
BDNF	Brain-derived neurotrophic factor
C	C-terminus
CAG	Cytosine, adenine, guanine triplet
CBD	Corticobasal degeneration
CP	20S core protease
DCTN1	Dynactin subunit 1
DCTN2	Dynactin subunit 2
GABA	Gamma-aminobutyric acid
GFAP	Glial fibrillary acidic protein
HAP1	Huntingtin-associated protein 1
HD	Huntington's disease
HSP	Heat shock protein
Htt	Huntingtin
IT15	Interesting transcript 15 gene or Huntington's disease gene
kDa	Kilodalton
LHE	Staining with Luxol fast blue and counterstaining with hematoxylin and eosin
MAP	Microtubule-associated protein
MRI	Magnetic resonance imaging
MSA	Multiple system atrophy
N	N-terminus
NDN	Neostriatal dark neurons
NES	Nuclear export signal
NI	Neuronal intranuclear inclusion
NPY	Neuropeptide Y

PD	Parkinson's disease
PSP	Progressive supranuclear palsy
RP	19S regulatory particle
SEP	Somatosensory evoked potentials
SCA1	Spinocerebellar ataxia type 1
SCA2	Spinocerebellar ataxia type 2
SCA3	Spinocerebellar ataxia type 3
SMI-32	Nonphosphorylated neurofilament proteins
UBA	Ubiquitin-associated domain
Ub	Ubiquitin
UbL	Ubiquitin-like domain
UPS	Ubiquitin-proteasome pathway
VEP	Visual evoked potentials

Chapter 1
Introduction

Since the late 1880s hereditary chorea is named Huntington's disease after *George Huntington* (1850–1916) (Fig. 1.1) who provided the classical description of adult onset of hereditary chorea in his essay "On chorea" from 1872. However, this publication was not the first clinical description of adult-onset hereditary chorea. It was

Fig. 1.1 The physician George Huntington (Reprinted from Lanska (2000), (Figure 2, page 83); with kind permission from Taylor & Francis)

U. Rüb et al., *The Neuropathology of Huntington's Disease: Classical Findings, Recent Developments and Correlation to Functional Neuroanatomy*,
Advances in Anatomy, Embryology and Cell Biology, Vol. 217,
DOI 10.1007/978-3-319-19285-7_1

preceded by earlier and fairly complete clinical descriptions (e.g., that of *Waters*, *Lund*, or *Lyon*) which already mentioned many of the key clinical features of the disease including its hereditary nature, insidious onset in adulthood, and progressive disease course (Heathfield 1973; Huntington 1872; Lanska 2000, 2010; Lund 1860; Lyon 1863; Waters 1842). In his essay from 1872, Huntington incorporated the medical records of the patients treated previously by his father and grandfather and noted the hereditary transmission of chorea, its gradual onset during adulthood and progressive course, the tendency of affected patients to insanity and suicide, as well as the resistance of the disease to treatments. In its main part this historical essay dealt with rheumatic or Sydenham's chorea, but its last part was devoted to a hereditary chorea with mental changes commencing in midlife, which is transmitted from generation to generation and follows a remorselessly progressive course (Heathfield 1973; Huntington 1872; Lanska 2000, 2010).

Widely believed to represent the initial descriptions of the brain pathology associated with Huntington's chorea and constituting milestones in neuropathological HD research, the neuropathological reports of *Gerbrandus Jelgersma* and *Alois Alzheimer* by no means constitute the first descriptions of the brain pathology in HD (Alzheimer 1911; Heathfield 1973; Jelgersma 1908; Lange and Aulich 1986).

From today's point of view, many of the historical neuropathological observations and descriptions revealed very valuable and pioneering postmortem findings. Although widely forgotten or neglected, many findings and assumptions of these early postmortem HD studies have been repeatedly confirmed during the following period of more than hundred years of neuropathological HD research. Certainly, the short historical review provided in the introduction of this monograph cannot do justice to all the involved HD researchers of the past, but it tries to capture the essential milestones of neuropathological HD research, to describe them in detail, and to explicitly appreciate their importance for the current neuropathological knowledge in the field of HD and its stepwise development during the last century. Owing to the large amount of neuropathological investigations performed worldwide during the last century, it is impossible to describe all findings of all these previous postmortem studies in this monograph in detail. Therefore, we have restricted our efforts to a collection of the most important discoveries of neuropathological HD research. With regard to the assumptions and confirmed, questionable, and unconfirmed postmortem findings of previous HD studies not mentioned here, we must refer to the original neuropathological literature.

Ewald Stier possibly was among the first neuroscientists who pointed to widespread neurodegenerative findings in the brains of patients with Huntington's chorea, which, astonishingly, are already compatible with the current view of HD as a multisystem neurodegenerative disease. Along with macroscopical brain alterations (i.e., reduced brain weight and brain atrophy), extensive degenerative changes were reported in subcortical brain regions (e.g., thalamus, putamen, and pallidum) as well as in circumscribed regions of the cerebral cortex (Stier 1903). In 1908, *Gerbrandus Jelgersma* described atrophic changes of the striatum (i.e., caudate nucleus and putamen), pallidum, and cerebral cortex in a patient with chronic chorea and attributed the occurrence of involuntary choreatic and incoordinated movements to these degenerative alterations in the striatum (see Chaps. 2, 3, and 4) (Jelgersma 1908).

Alois Alzheimer (Fig. 1.2) apparently also recognized the widespread brain pathology in Huntington's chorea in 1911. Along with degeneration of the striatum and pallidum, he mentioned the involvement of additional subcortical regions (e.g., thalamus, brainstem, and spinal cord) and the cerebral cortex and attributed the occurrence of chorea to the degeneration of the striatum (see Chaps. 2, 3, 4, and 7). Currently representing a well-known and universally accepted, characteristic histopathologic feature of HD, the differential vulnerability of small and large striatal nerve cells for the degenerative process of HD has been delineated by *James Ramsey Hunt* and among others has been confirmed by *Cécile and Oskar Vogt* in their classical, comprehensive work on the human striatum and its diseases (see Chaps. 2 and 4) (Hunt 1917; Vogt and Vogt 1920). In their seminal work Cécile and Oskar Vogt (Fig. 1.3) mentioned, apart from degeneration of the striatum and pallidum, an involvement of the claustrum, subthalamic nucleus, and spinal cord in chorea patients, as well as a widespread loss of cortical pyramidal cells and preservation of the giant Betz pyramidal cells in the primary motor cortex (see Chaps. 2, 3, 4, and 11) (Vogt and Vogt 1920). Subsequently *Friedrich H. Lewy* described an additional affection of the substantia nigra and cerebellum in chorea patients (see Chaps. 5, 6, and 7) (Lewy 1923). Early reports of a possible involvement of gray components of the medial temporal lobe in chorea patients are scanty and pointed to degenerative

Fig. 1.2 The physician and neuropathologist Alois Alzheimer (Reprinted from Jellinger (2006), (Figure 1, page 114); with kind permission from Springer Science and Business Media)

Fig. 1.3 The neuroscientists Cécile and Oskar Vogt (Reprinted from Jones (2003), (Figure, page 19); with kind permission from Nature Publishing Group)

changes in the allocortical hippocampus (Terplan 1924) and amygdala (see Chaps. 3 and 11) (Davison et al. 1932).

Although traditionally regarded as a brain disease which is more or less confined to the striatum, many of the very early neuropathological observations already gave rise to doubts about the validity of the reductionistic pathoanatomical and pathophysiological HD concepts, which were unilaterally based on the well-known affection of the striatum. Numerous qualitative and quantitative studies of the past indicated that the extent of brain neurodegeneration in HD has been underestimated for a long time, is more widespread and severe than originally believed, goes far beyond the well-known lesions in the striatum (i.e., caudate nucleus and putamen), and also involves the cerebral cortex, various additional subcortical regions, as well as the cerebellum (see Chaps. 2, 3, 4, 5, 6, 7, and 11). The stepwise scientific progress in the neuropathology of HD made during a time period of more than one century eventually led to the replacement of the traditional pathoanatomical and pathophysiological HD concepts and culminated in the current concept of HD as a *multisystem degenerative disease* of the human brain (Heinsen et al. 1994; Lange and Aulich 1986; Rüb et al. 2013a; Vonsattel 2008). This concept constitutes a reliable base for further studies aimed at elucidation of the pathophysiological and pathobiochemical mechanisms leading to neurodegeneration and may thus pave the way for the development of effective, causative, and/or disease-modifying therapies for affected HD patients (Fig. 1.4).

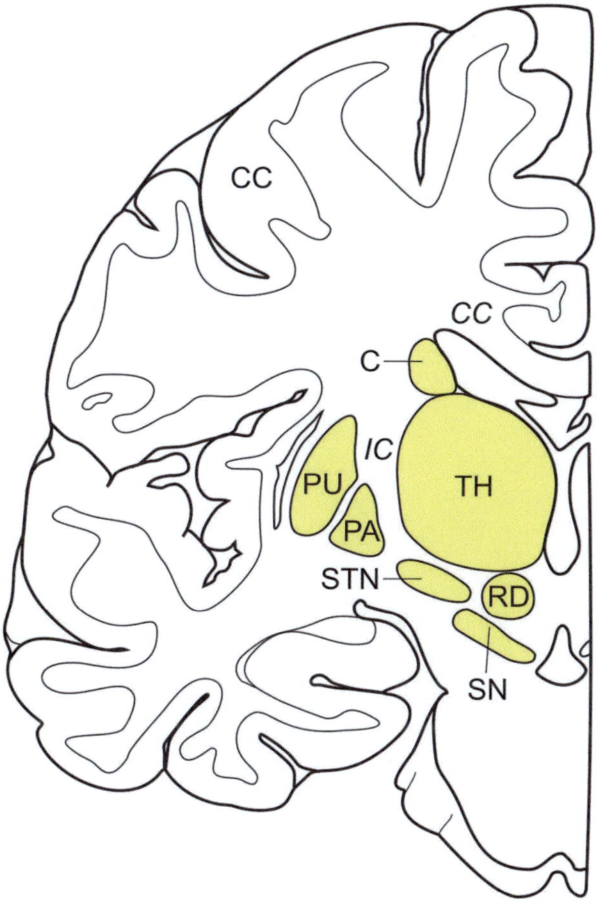

Fig. 1.4 Distribution of volume and neuronal loss in select brain regions of Huntington's disease (HD) patients. Schematized frontal section through the cerebral hemisphere and midbrain of the human brain at the level of the red nucleus (*RD*). Recent detailed postmortem studies using quantitative morphometric methods detected volume loss in the following brain components of Huntington's disease (HD) patients: accumbens nucleus, 47 %; amygdala, 26 %; cerebral cortex (*CC*), 15 %; cerebral white matter, 23 %; caudate nucleus (*C*), 52 %; putamen (*PU*), 57 %; pallidum (*PA*), 50 % (external segment), 40 % (internal segment); thalamus (*TH*), 14 %; subthalamic nucleus (*STN*), 17 %; substantia nigra (*SN*), 40 % (compact part), 60 % (reticulate part); red nucleus (12 %). Furthermore estimations of nerve cell numbers revealed neuronal loss in the following structures: *C*, 86 % (small nerve cells), 41 % (large nerve cells); *PU*, 85 % (small nerve cells), 57 % (large nerve cells); *PA*, 26 % (external segment), 33 % (internal segment); *STN*, 17 % (Lange and Aulich 1986). Brain structures that may undergo degeneration during HD are colored in *yellow.* Abbreviations: *C* caudate nucleus, *CC* cerebral cortex, *CC* corpus callosum, *IC* internal capsule, *PA* pallidum, *PU* putamen, *RD* red nucleus, *SN* substantia nigra, *STN* subthalamic nucleus, *TH* thalamus

This chapter was written by Udo Rüb.

Chapter 2
The Neuropathological Grading of Huntington's Disease (HD)

2.1 Neuropathological Base for the Grading System of Huntington's Disease (HD)

The deleterious action of the unstable CAG repeat expansion in the *Huntington's disease* (*HD*) *gene* (also called *IT15*) located on the short arm of chromosome 4 involves widespread areas of the brain including sites of increased vulnerability or sites that are relatively resistant, but not spared (Andrew et al. 1993; Duyao et al. 1993; Myers et al. 1991; The Huntington's disease Collaborative Research Group 1993). Therefore, the intensity of the degenerative process of HD is topographically variable. The neuropathological phenotype of adult HD can be almost cryptic or outstanding within the same nucleus, but at different sites. Indeed, the expression of the degenerative process differs not only among distinct anatomical compartments, but also within specific brain compartments (e.g., cerebral cortex, white matter, striatum, pallidum, thalamus, brainstem, cerebellum), or systems (e.g., basal ganglia, limbic system) (Fig. 1.4) (Braak and Braak 1992a, b; Bruyn et al. 1979; De la Monte et al. 1988; Dom et al. 1976; Dunlap, 1927; Duyao et al. 1993; Estrada-Sanchez and Rebec 2013; Fennema-Notestine et al. 2004; Ferrante et al. 1987; Hedreen et al. 1991; Heinsen et al. 1992, 1994, 1996, 1999; Heinsen and Rüb 1997; Lange 1981; Lange and Aulich 1986; Lange et al. 1976; Myers et al. 1988; Rüb et al. 2013a, 2014a, b; Selemon et al. 2004; Shoulson and Young 2011; Sotrel et al. 1991; Vogt and Vogt 1920, 1942; Vonsattel 2008; Vonsattel and DiFiglia 1998; Vonsattel et al. 1985). Furthermore, the expression of the pathological phenotypes depends on a constellation of influences driven mainly by epigenetic factors, genetic modifiers, duration of symptoms, or the idiosyncratic longevity of the patients (Hodges et al. 2006; U.S.-Venezuela Collaborative Research Project and Wexler 2004). Thus, upon postmortem examination the pathological phenotypes of HD brains are more or less obvious depending on the sites or systems that are

U. Rüb et al., *The Neuropathology of Huntington's Disease: Classical Findings, Recent Developments and Correlation to Functional Neuroanatomy*,
Advances in Anatomy, Embryology and Cell Biology, Vol. 217,
DOI 10.1007/978-3-319-19285-7_2

considered and on the techniques applied for assessing the brains. The involvement and the evolution of the neurodegenerative changes in the *striatum* (caudate nucleus, putamen), and pallidum (paleostriatum) strikingly underscore the differential regional vulnerability occurring in HD within this discrete, relatively small subregions of the brain (Fig. 1.4) (Birnbaum 1941; Braak and Braak 1992a, b; Bruyn et al. 1979; De la Monte et al. 1988; Dom et al. 1976; Dunlap 1927; Estrada-Sanchez and Rebec 2013; Fennema-Notestine et al. 2004; Ferrante et al. 1987; Forno and Jose 1973; Hedreen et al. 1991; Heinsen et al. 1992, 1994, 1996, 1999; Heinsen and Rüb 1997; Hodges et al. 2006; Kiesselbach 1914; Landwehrmeyer et al. 1995; Lange 1981; Lange and Aulich 1986; Lange et al. 1976; Lewy 1923; McCaughey 1961; Myers et al. 1988; Neustaedter 1933; Roos et al. 1985; Rüb et al. 2013a, b, 2014a, b; Schroeder 1931; Selemon et al. 2004; Sotrel et al. 1991; Terplan 1924; Vonsattel 2008; Vonsattel and DiFiglia 1998; Vonsattel et al. 1985).

The *vulnerability of the striatum* correlates with the size of the HD-IT15 CAG repeat expansion (Furtado et al. 1996; Penney et al. 1997). The primary HD-related atrophy might be exacerbated in some instances by the age-related involution of the brain notably in patients with the lower end of the HD-IT15 CAG repeat expansion or with a concomitant Alzheimer's disease (AD)-related brain pathology. The mechanisms causing this increased vulnerability of the striatum for the pathological process of HD are unknown.

Because the striatum of HD patients shows regional variation of the severity and progression of the degenerative process, a *grading of the HD-associated striatal neuropathology* was developed. This grading is based on the fact that the gradual atrophy of the neostriatum which is due to the slowly ongoing loss of neurons and the concomitant occurrence of reactive astrogliosis with differential regional expression over time has been for a long time regarded as the major neuropathological hallmark of HD.

The aim of grading the neuropathological severity is to sort and to pool samples of the striata of HD patients for research, which share comparable "Gestalt" patterns of the parenchyma. Indeed, the grading identifies striatal samples with subregions that are relatively spared (e.g., accumbens nucleus), or subregions in which the neuronal loss and reactive astrocytosis are moderate, or subregions in which the residual cells mainly consist of reactive astrocytes, oligodendrocytes, and microglial cells. Investigations focusing on the reasons for the vulnerability of the HD striatum increasingly require semiquantitative measures of the degenerative process, which are also useful for studies correlating striatal changes with changes outside of the striatum in the brains of HD patients or for clinicopathological correlations. The reproducible, rather simple grading system is widely used for studies focusing on the regional and cellular susceptibility to the toxic effects of the HD-IT15 CAG repeat expansion, which are operating directly within or indirectly outside the striatum. To some extent this grading system allows for a selection of brain samples for research with a reasonable estimation of the type of residual cells they contain (e.g., mixed glial and neuronal populations or an overwhelming glial population) (Vonsattel 2008; Vonsattel and DiFiglia 1998; Vonsattel et al. 1985).

The neuropathological HD grading provides these semiquantitative measures of the striatum at both the macroscopical (gross) and microscopical levels. Grossly the striatum is assessed by the use of three standardized thick frontal slices through the cerebral hemispheres (Figs. 2.1, 2.2, 2.3, 2.5, and 2.6), while microscopical evaluation is performed on tissue sections through four different levels of the striatum of HD patients (Figs. 2.2, 2.3, 2.4 2.5, 2.6, 2.7, and 2.8).

2.2 Grading of Striatal Neuropathology in Huntington's Disease (HD)

The *striatal neurodegeneration of HD* has an ordered and topographic evolution and distribution. Briefly, along the sagittal axis of the brain, the tail of the caudate nucleus shows more degeneration than the body of the caudate nucleus, which in turn is more involved than the head of the caudate nucleus. Likewise, the caudal portion of the putamen is more degenerated than the rostral portion. Along the coronal axis of the brain, the dorsal regions of the striatum are more involved than the ventral ones and the medial ones more than the lateral ones. Thus, from a three-dimensional point of view, the worsening of the degenerative process in the striatum, which encompasses neuronal loss and reactive astrogliosis, appears to simultaneously move in a caudo-rostral direction and concomitantly, in both, the dorsoventral and mediolateral directions. Gradually a regional, differential, and slowly decreasing density of neostriatal neurons becomes inversely proportional to the density and severity of the slowly emerging reactive astrogliosis, which parallels the severity of the neuronal loss. Most residual neurons in the striatum have normal somatic morphology and increased lipofuscin and are perhaps smaller than expected. Among these apparently normal striatal nerve cells, there are scattered atrophic neurons that appear more darkly stained in tissue sections treated with Luxol fast blue and counterstained with hematoxylin and eosin (LHE). These atrophic neurons are referred to as *neostriatal dark neurons* (*NDN*) and have a distinct scalloped cell membrane, a granular dark cytoplasm, and an oblong nucleus with dense chromatin. NDN tend to form ill-defined clusters and are scarce in both the atrophic and the relatively spared neostriatal zones. Their density tends to increase in the intermediary zone of the neostriatum, which is flanked by the two other zones (i.e., the severely involved dorsal zone and the relatively spared ventral zone) (Vonsattel 2008; Vonsattel and DiFiglia 1998; Vonsattel et al. 1985).

This neuropathological classification system for HD has *5 grades* (*0 – 4*) *of the severity and evolution of the striatal involvement*. The assignment of a grade of neuropathological HD severity depends on the extent of the striatal atrophy (*macroscopical examination*) and on the extent of the striatal neuronal loss and degree of reactive astrogliosis (*microscopical examination*) (Vonsattel 2008; Vonsattel and DiFiglia 1998; Vonsattel et al. 1985).

Gross examination is performed on three standardized 0.3–0.5 cm-thick frontal brain sections that include four distinct levels of the striatum: (1) *level CAP* with the

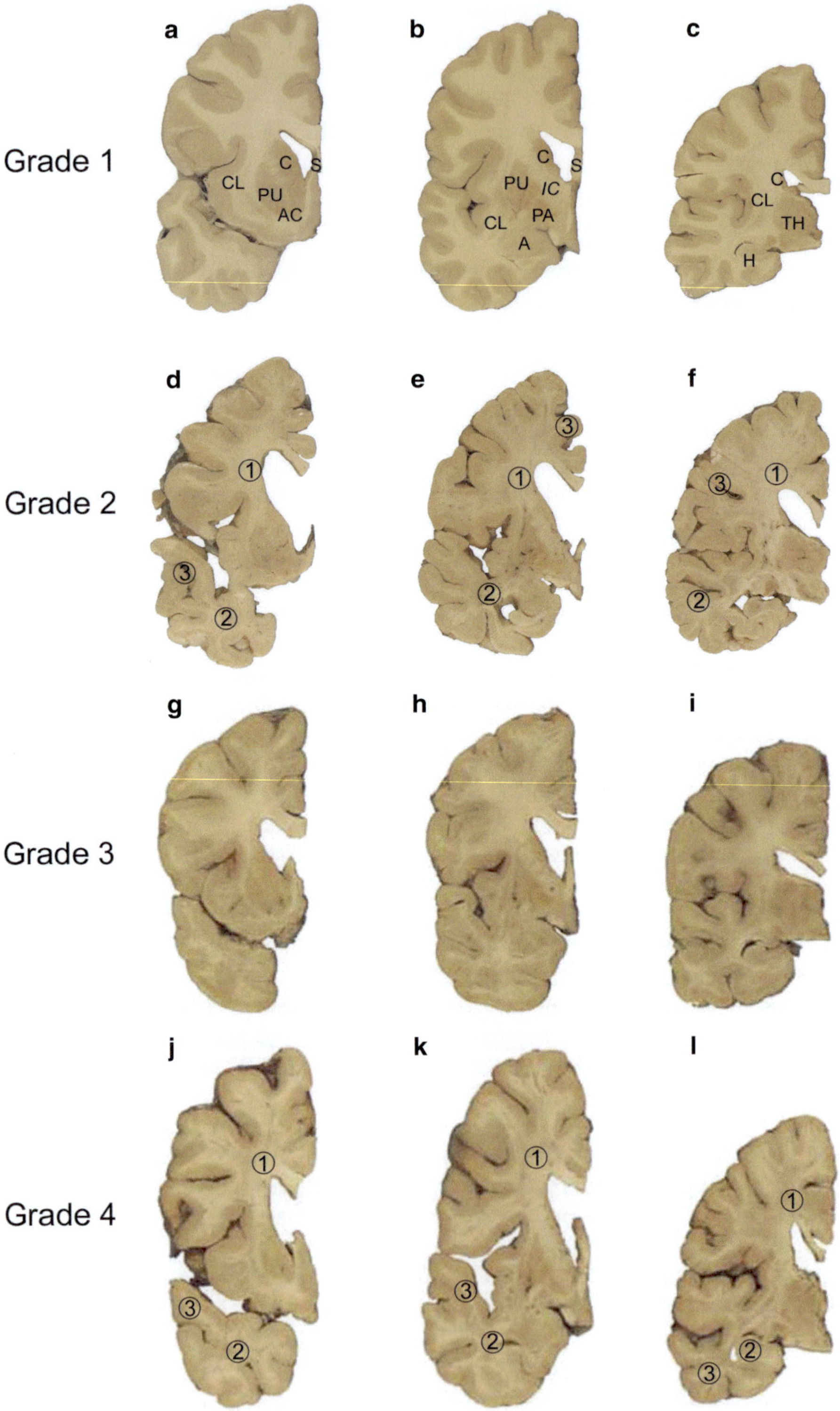
Grade 1
Grade 2
Grade 3
Grade 4
a
b
c
d
e
f
g
h
i
j
k
l
C
S
CL
PU
AC
IC
PA
A
TH
H
①
②
③

Fig. 2.1 Neuropathological grading of Huntington's disease (HD). (**a–c**) *Grade 1 of neostriatal atrophy in HD*. Thick frontal sections through the brain of a 29-year-old male Huntington's disease (HD) patient. The male HD patient received the mutated HD gene from his mother, suffered from motor clumsiness and gait changes, showed aggressive behavior, and died from the results of a fatal accident. (**a**) Rostral caudate nucleus (*C*), claustrum (*CL*), putamen (*PU*), and accumbens nucleus (*AC*) (*level CAP*). (**b**) Pallidum (*PA*) with the caudal portion of the head of the caudate nucleus and the adjacent putamen (*level GP*). (**c**) Lateral geniculate body of the thalamus (*TH*) with the body (dorsal) and tail of the caudate nucleus (ventral) (*level caudal thalamus*). *Macroscopical changes* of the brain of the patient corresponding to *grade 1 of neostriatal atrophy*: Atrophy of the body of the caudate nucleus and reduction of its tail (**b**, **c**). Normal gross aspect of the putamen and accumbens nucleus (**a–c**). (**d–f**) *Grade 2 of neostriatal atrophy in HD*. Thick frontal sections through the brain of an 89-year-old male HD patient with 17 CAG trinucleotide repeats in the normal HD gene and 40 CAG repeats in the mutated HD gene. This HD patient received the mutated HD gene from his mother HD. HD onset in this patient was at age 75, and definite involuntary choreatic movements first appeared at age 79. (**d**) Rostral caudate nucleus, putamen, and accumbens nucleus (*level CAP*). (**e**) Pallidum with the caudal portion of the head of the caudate nucleus and the adjacent putamen (*level GP*). (**f**) Lateral geniculate body of the thalamus with the body (dorsal) and tail of the caudate nucleus (ventral) (*level caudal thalamus*). *Macroscopical brain changes* corresponding to *grade 2 of neostriatal atrophy*: evident gross striatal atrophy and enlarged lateral ventricles (**d–f**). Despite marked atrophy of the head of the caudate nucleus, the convex outline of its ventricular surface is still retained (**d**). The involvement of the body and tail of the caudate nucleus is still discrete (**e**, **f**). The putamen and pallidum show an obvious volume loss (**d**, **e**), while the accumbens nucleus appears macroscopically normal (**d**). Some of the non-striatal atrophic changes of the brain of the patient are in part due to a concomitant Alzheimer's disease (AD)-related pathology. Note the additional (*1*) cerebral white matter loss, (*2*) atrophy of the temporal lobe, and (*3*) reduction of the gray band of the cerebral cortex (**d–f**). (**g–i**) *Grade 3 of neostriatal atrophy in HD*. Thick frontal sections through the brain of a 40-year-old female HD patient with 17 CAG trinucleotide repeats in the normal HD gene and 49 CAG repeats in the mutated HD gene. HD onset was at age 32. The patient received the gene via paternal transmission. (**g**) Rostral caudate nucleus, putamen, and accumbens nucleus (*level CAP*). (**h**) Pallidum with the caudal portion of the head of the caudate nucleus and the adjacent putamen (*level GP*). (**i**) Lateral geniculate body of the thalamus with the body (dorsal) and tail of the caudate nucleus (ventral) (*level caudal thalamus*). *Macroscopical brain changes* corresponding to *grade 3 of neostriatal atrophy*: severe atrophy of the caudate nucleus with shrunken head, barely distinguishable body (dorsal) and tail (ventral) (**g–i**). The medial outline of the caudate nucleus forms a straight-line configuration, which parallels the anterior limb of internal capsule (**g–i**). The accumbens nucleus still has a normal appearance (**g**). (**j–l**) *Grade 4 of neostriatal atrophy in HD*. Thick frontal sections through the brain of a 47-year-old female HD patient with 17 CAG trinucleotide repeats in the normal HD gene and 54 CAG repeats in the mutated HD gene. HD onset was at age 29. (**j**) Rostral caudate nucleus, putamen, and accumbens nucleus (*level CAP*). (**k**) Pallidum with the caudal portion of the head of the caudate nucleus and the adjacent putamen (*level GP*). (**l**) Lateral geniculate body of the thalamus with the body (dorsal) and tail of the caudate nucleus (ventral) (*level caudal thalamus*). *Macroscopical brain changes* corresponding to *grade 4 of neostriatal atrophy*: severely atrophic neostriatal caudate nucleus and putamen with markedly concave medial outlines of the head of the caudate nucleus and adjacent internal capsule (**j–l**). In HD grade 4 individuals, neostriatal neuronal loss commonly amounts to 95 % and leaves behind a vacuolated neuropil in most affected regions of the neostriatum. The pallidum likewise is severely degenerated and sustains neuronal loss. Although remaining relatively preserved in at least 50 % in individuals with striatal grade 4 neuropathology, the accumbens nucleus in these individuals is by no means spared. Note the additional severe (*1*) degeneration of the cerebral white matter, (*2*) atrophy of the temporal lobe, and (*3*) reduction of the gray band of the cerebral cortex (**j–l**). Abbreviations: *A* amygdala, *AC* accumbens nucleus, *C* caudate nucleus, *CL* claustrum, *H* hippocampus, *IC* internal capsule, *PA* pallidum, *PU* putamen, *S* septum, *TH* thalamus

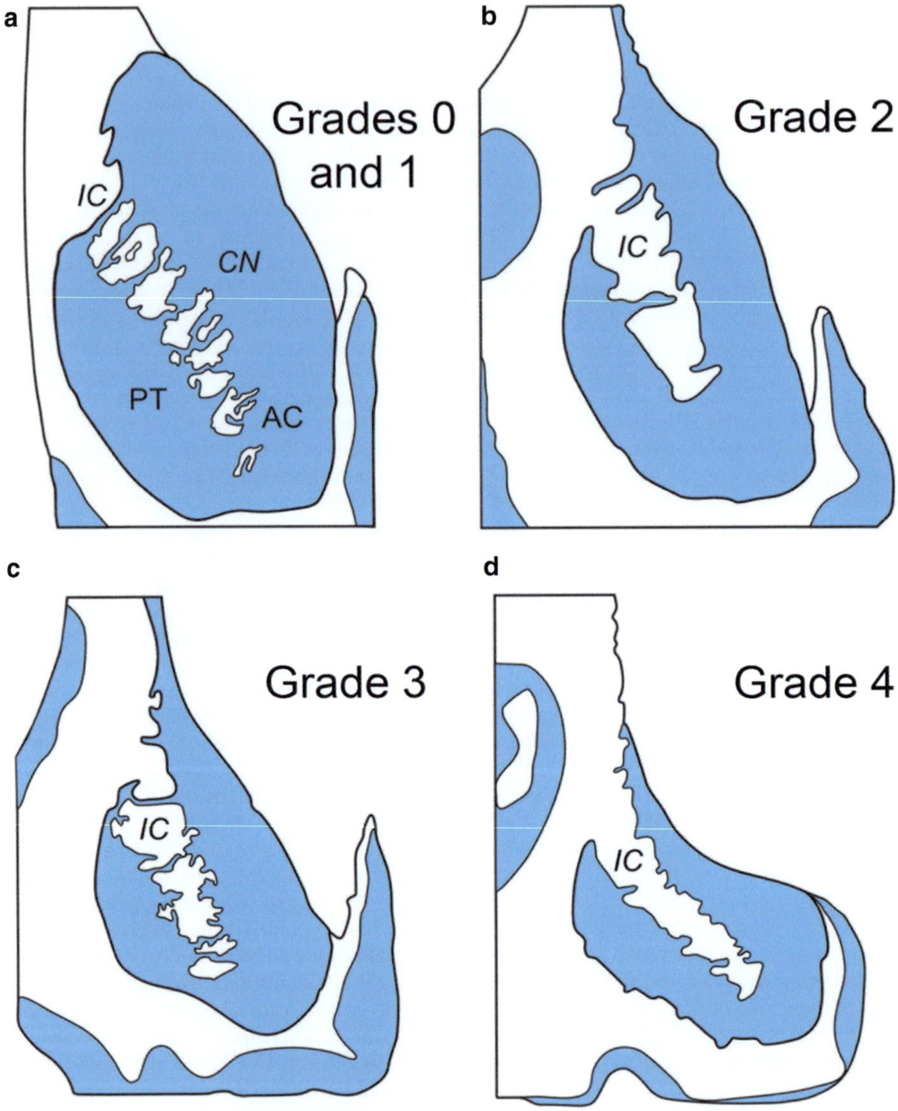

Fig. 2.2 Progressive degeneration of the striatum in Huntington's disease (HD). (**a–d**) Schematized frontal sections through the rostral level of the striatum with the head of the caudate nucleus (*CN*), putamen (*PT*), and accumbens nucleus (*AC*) (*level CAP*). (**a**) Grades 0 and 1 of striatal atrophy in Huntington's disease (HD): normal gross aspect of the neostriatum (CN and PT) and AC. (**b**) Grade 2 of striatal atrophy in HD: markedly atrophic head of the CN. The convex outline of the ventricular surface of the CN, however, is still retained. The PT already shows an obvious volume loss, while the AC appears macroscopically normal. (**c**) Grade 3 of striatal atrophy in HD: moderate to severe atrophy of the neostriatum. The medial outline of the CN is now flat and forms a nearly straight-line configuration, which parallels the anterior limb of the internal capsule (*IC*). (**d**) Grade 4 of striatal atrophy in HD: very severe atrophic CN and PT with markedly concave medial outline of the head of the CN and *IC*. The AC of grade 4 individuals is also atrophic (Modified according to Vonsattel et al. (1985), (Figure 2, page 566); with kind permission from Wolters Kluwer Health). Abbreviations: *AC* accumbens nucleus, *CN* caudate nucleus, *IC* internal capsule, *PT* putamen

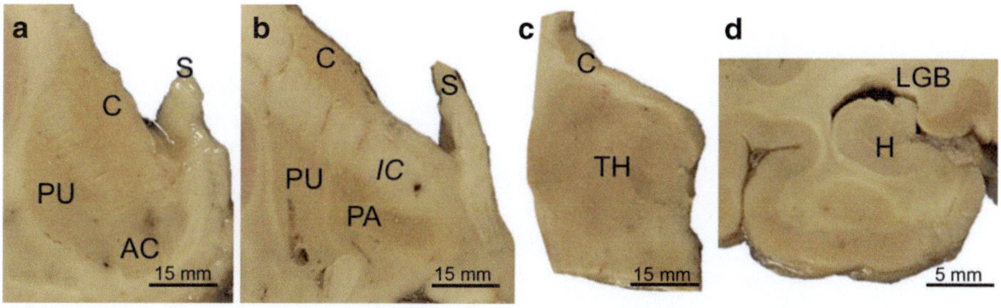

Fig. 2.3 Standardized brain tissue blocks for the macroscopical evaluation of the grade of striatal atrophy in Huntington's disease (HD). The assignment of a grade of the severity of the striatal pathology in the brains of Huntington's disease (HD) patients is based on macroscopical and microscopical investigations of the striatum. The *macroscopical investigation* includes three standardized thick frontal brain slices that comprise four distinct levels of the striatum: (1) *CAP level* with the caudate nucleus, accumbens nucleus, and putamen (see Fig. 2.1a, d, g, j). (2) *GP level* with the pallidum and caudal portion of the head of caudate nucleus, as well as the adjacent putamen (see Fig. 2.1b, e, h, k). (3) *Caudal level of the thalamus* with the lateral geniculate body, as well as the body of the caudate nucleus (see Fig. 2.1c, f, i, l). For the *microscopical examination*, four blocks are obtained from the following thick frontal slices: (**a**) *CAP level*. (**b**) *GP level*. (**c**) *Caudal level of thalamus* through the centromedian-parafascicular complex and the body of the caudate nucleus. (**d**) *Level of the hippocampal formation* at the thalamic lateral geniculate body with the tail of the caudate nucleus. Figures (**a–d**) are taken from the 40-year-old female HD grade 3 patient of Fig. 2.1. Abbreviations: *AC* accumbens nucleus, *C* caudate nucleus, *H* hippocampus, *IC* internal capsule, *LGB* lateral geniculate body of the thalamus, *PA* pallidum, *PU* putamen, *S* septum, *TH* thalamus

head of the caudate nucleus, accumbens nucleus, and putamen (Figs. 2.1a, d, g, j and 2.2); (2) *level GP* with the pallidum, the caudal portion of the head of the caudate nucleus, and the adjacent putamen (Figs. 2.1b, e, h, k and 2.2); and (3) *level caudal thalamus* through the lateral geniculate body with the body and tail of the caudate nucleus (Figs. 2.1c, f, i, l and 2.2).

For the *microscopical examination* of the distribution and the extent of the HD-related neuronal loss and reactive astrogliosis, four tissue blocks are obtained from these four thick frontal hemispheric sections: (1) level CAP, (2) level GP, (3) level caudal thalamus, and (4) level of hippocampal formation at the lateral geniculate body (Figs. 2.2, 2.3, and 2.4). The techniques applied for the microscopical evaluation are those conventionally used in the setting of clinical neuropathology laboratories.

Among more than 1.250 HD brains investigated in our laboratory so far, less than one percent showed "grade 0," four percent "grade 1," 16 % "grade 2," 52 % "grade 3," and 28 % "grade 4" striatal neuropathology.

Grade 0 Grade 0 is assigned to any brain of carriers of the HD-IT15 CAG repeat expansion if, on *gross examination*, the morphological brain features are indistinguishable from those of healthy individuals (Figs. 2.2 and 2.5). Qualitative *microscopical investigation* of tissue sections treated with LHE reveals no reliable striatal neuronal loss or reactive astrogliosis (Fig. 2.4a, e). However, further evaluations including

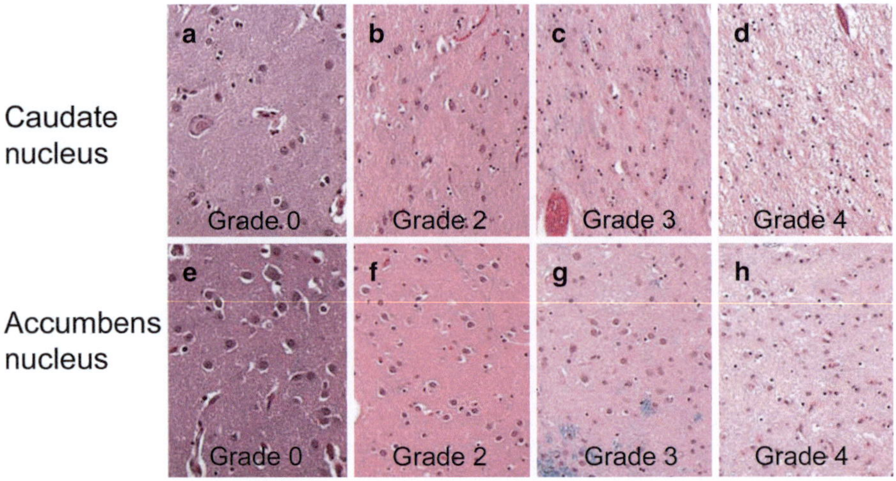

Fig. 2.4 Microscopical changes in the striatum in Huntington's disease (HD). (**a**, **b**, **c**, **d**) Luxol fast blue treated tissue sections counterstained with hematoxylin and eosin (LHE) through the head of the caudate nucleus (dorsal third at midpoint between the ependyma and the medial edge of the anterior limb of the internal capsule). (**e**, **f**, **g**, **h**) LHE treated tissue sections through the accumbens nucleus (ventral third, at midpoint along its horizontal axis) (magnification: 400×). (**a**) Regular nerve cell density in the head of the caudate nucleus and (**e**) accumbens nucleus of an 80-year-old male control individual. (**b**) Definite nerve cell loss and reactive astrogliosis in the head of the caudate nucleus and (**f**) unremarkable accumbens nucleus of an 89-year-old male Huntington's disease (HD) patient with grade 2 of striatal atrophy (see also Fig. 2.1d–f). (**c**) Severe neuronal loss and reactive astrogliosis in the head of the caudate nucleus and (**g**) not more than mildly decreased neuronal density in the accumbens nucleus of a 40-year-old female HD patient with grade 3 striatal atrophy (see also Fig. 2.1g–i). (**d**) Nearly complete neuronal loss, vacuolated neuropil, and reactive astrogliosis in the head of the caudate nucleus and (**h**) decreased neuronal density in the accumbens nucleus of a 47-year-old female HD patient with grade 4 striatal atrophy (see also Fig. 2.1j–l)

quantitative methods for cell counting demonstrate a 30–40 % loss of neurons and the absence of a visible reactive astrocytosis in the head of the caudate nucleus. In addition, in presymptomatic HD gene carriers with grade 0 striatal neuropathology ubiquitinated *neuronal nuclear inclusions*, and an increased density of oligodendrocytes may be present in the tail of the caudate nucleus (Fig. 2.5) (Gómez-Tortosa et al. 2001). The neocortical and neostriatal neuronal nuclear inclusions may occur long before any other neuropathological changes can be detected with conventional methods: they can be visualized with immunocytochemical techniques using antibodies directed against the degradation-related protein ubiquitin or with antibodies raised against pathologically expanded polyglutamine tracts. In summary, although apparently normal, grade 0 brains are most likely harboring changes that are not detectable with the conventional neuropathological methods. Unless methods are applied that are beyond the ones conventionally used in the clinical neuropathology setting, the HD-related tissue changes in individuals with grade 0 striatal neuropathology would remain cryptic (Vonsattel 2008; Vonsattel and DiFiglia 1998; Vonsattel et al. 1985).

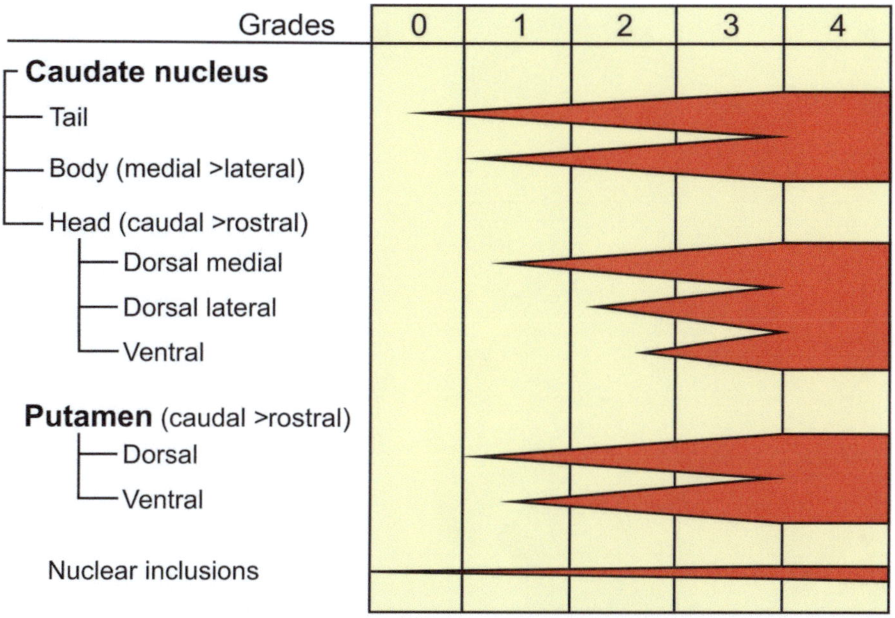

Fig. 2.5 Progression of striatal neuronal loss and astrogliosis in Huntington's disease (HD). Schematic diagram summarizing the neuropathological grades and the topographical variation and progression of neostriatal neuronal loss and astrogliosis characteristic of Huntington's disease (HD). The gradients of neuronal loss decrease along the caudo-rostral, mediolateral, and dorsoventral axes of the neostriatum, while the gradients of the reactive astrogliosis increase along these neostriatal axes. Owing to the extent of degeneration, these neuropathological gradients are often blurred in grade 4. Neuronal nuclear inclusions of the disease protein huntingtin (HTT) may evolve already in the absence of striatal neurodegeneration (grade 0) and their formation progresses during the neuropathological HD grades 1–4 (Modified according to Vonsattel and DiFiglia (1998), (Figure 6, page 375); with kind permission from Wolters Kluwer Health)

Grade 1 In grade 1 the body and head of the caudate nucleus and the putamen may appear normal *on gross examination* (Fig. 2.2). In contrast the tail of the caudate nucleus is much smaller than normal, and considerable atrophy is also noticeable in the body of the caudate nucleus (Figs. 2.1a–c and 2.5). *Neuronal loss and astrogliosis* in grade 1 individuals for most involve the tail of the caudate nucleus and, less severely, the body of the caudate nucleus, the dorsal subregions of the head of the caudate nucleus, and the nearby putamen. While the accumbens nucleus is unremarkable, cell counts in grade 1 individuals commonly reveal neuronal loss of approximately 50 % or more in the head of the caudate nucleus (Figs. 2.2 and 2.5) (Vonsattel 2008; Vonsattel and DiFiglia 1998; Vonsattel et al. 1985).

Grade 2 In grade 2 the lateral ventricles are enlarged. Although *gross striatal atrophy* is evident, the convex outline of the ventricular surface of the head of the caudate nucleus is still retained (Figs. 2.1d–f, 2.2, and 2.5). Atrophy of the caudate

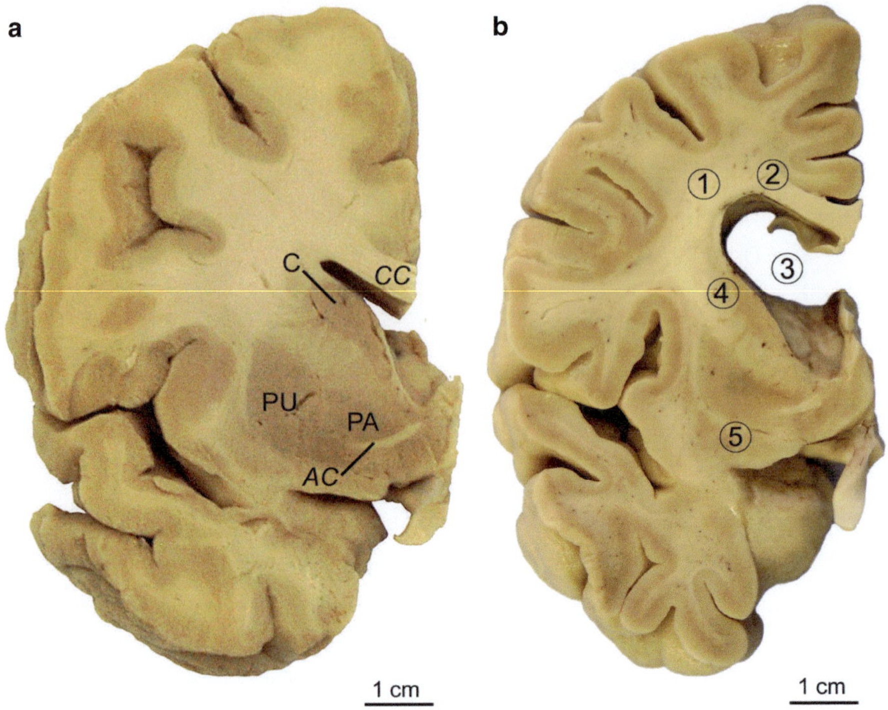

Fig. 2.6 Atrophy of the striatum and white matter loss in Huntington's disease (HD). (**a**) Frontal section through the basal forebrain of a representative control individual. (**b**) Frontal section through the left basal forebrain of a clinically diagnosed and genetically confirmed Huntington's disease (HD) patient. Note the loss of deep white matter (*1*), the narrowed corpus callosum (*CC*) (*2*), and widened third ventricle (*3*). The atrophy of the caudate nucleus (*C*) (*4*) and putamen (*PU*) (*5*) corresponds to Vonsattel grade 2 of neostriatal atrophy (Reprinted from Rüb et al. (2009), (Figure 2, page 6); with kind permission from John Wiley and Sons). Abbreviations: *AC* anterior commissure, *C* caudate nucleus, *CC* corpus callosum, *PA* pallidum, *PU* putamen

nucleus is more conspicuous at the GP level (Figs. 2.1e and 2.5) than at the CAP level (Figs. 2.1d and 2.5), and atrophy of the body and tail of the caudate nucleus is discrete (Figs. 2.1f and 2.5). In grade 2 volume loss of the lenticular nucleus (putamen and pallidum) is mild on general survey (Figs. 2.1e, 2.2, and 2.5). In grade 2 *definite neuronal loss and reactive gliosis* are seen in the dorsal half (medial > lateral) of the caudate nucleus (Figs. 2.2, 2.4b, and 2.5) and dorsal adjacent putamen. The accumbens nucleus appears microscopically normal (Figs. 2.1, 2.2, and 2.5) (Vonsattel 2008; Vonsattel and DiFiglia 1998; Vonsattel et al. 1985).

Grade 3 In grade 3 the head of the caudate nucleus is shrunken, and its medial outline forms a straight-line configuration, which parallels the anterior limb of the internal capsule (Figs. 2.1g–i, 2.2, and 2.5). At level CAP and beyond it, the caudate nucleus is severely atrophic, and its body and tail are barely distinguishable grossly (Figs. 2.1g–i, 2.2, and 2.5). Macroscopically, the accumbens nucleus

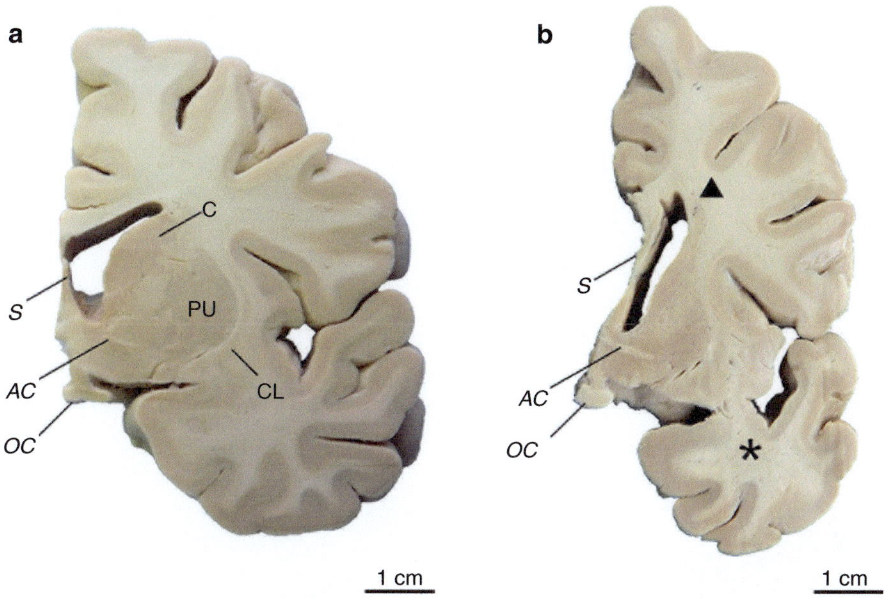

Fig. 2.7 Degeneration of the striatum and white matter loss in Huntington's disease (HD). (**a**) Frontal section through the basal forebrain of a representative control individual at the level of the optic chiasm (*OC*) with the caudate nucleus (*C*) and putamen (*PU*). (**b**) Frontal section through the same level of the left basal forebrain of a clinically diagnosed and genetically confirmed male Huntington's disease (HD) patient. The atrophy of the C and PU of this HD patient corresponds to Vonsattel grade 2 of neostriatal atrophy. Note the additional atrophy of the temporal lobe (*asterisk*) and loss of deep white matter (*triangle*). Abbreviations: *AC* anterior commissure, *C* caudate nucleus, *CL* claustrum, *PU* putamen, *OC* optic chiasm, *S* septum

appears unremarkable (Figs. 2.1g, 2.2, and 2.5). In grade 3 individuals *neuronal loss and fibrillary astrocytosis* are severe within the caudate nucleus and putamen (rostral > caudal; dorsal > ventral) (Figs. 2.4c and 2.5), whereas the neuronal density in the accumbens nucleus is either apparently normal or only mildly decreased (Figs. 2.2, 2.4g, and 2.5) (Vonsattel 2008; Vonsattel and DiFiglia 1998; Vonsattel et al. 1985).

Grade 4 In grade 4 the *striatum* is severely atrophic. Along with that of the internal capsule, the medial contour of the head of the caudate nucleus is concave (Figs. 2.1j–l, 2.2, and 2.5). In addition, the volume of the pallidum is reduced by about 50 %. The accumbens nucleus remains relatively preserved, however, not spared in at least 50 % of grade 4 individuals (Figs. 1.4, 2.1j, 2.2, 2.4h, and 2.5). The *neostriatal neuronal loss* in grade 4 amounts to 95 % or more and leaves behind a vacuolated neuropil (Figs. 1.4, 2.4d, and 2.5) (Vonsattel 2008; Vonsattel and DiFiglia 1998; Vonsattel et al. 1985).

Bilateral atrophy of the neostriatum (Figs. 1.4, 2.1g–l, 2.2, 2.3, 2.4, 2.5, 2.6, 2.7, and 2.8) is invariably found in the late clinical stages of Huntington's disease (HD), represents the most prominent macroscopical neuropathological finding, and is

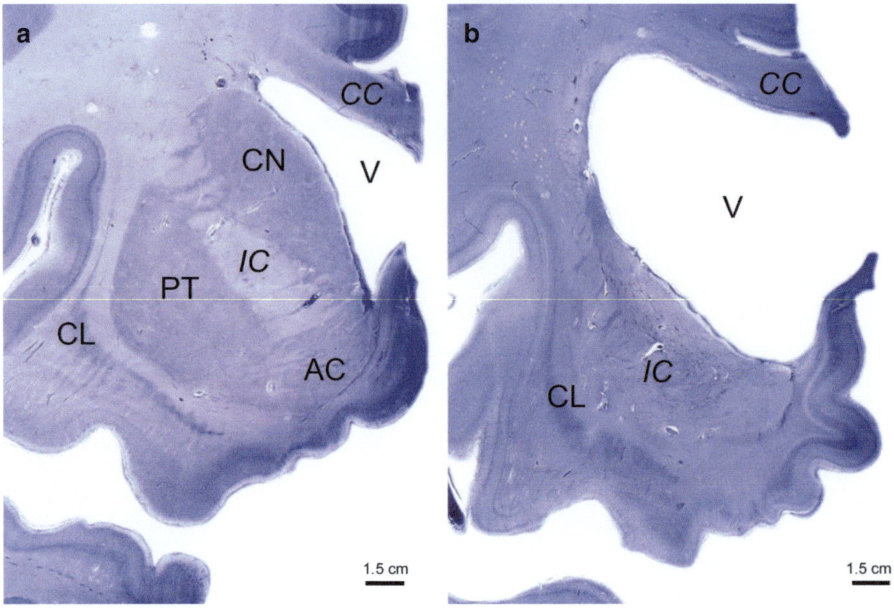

Fig. 2.8 Neuronal loss in the neostriatum in Huntington's disease (HD). (**a**) Frontal section through the rostral portion of the left striatum of a representative control individual with the neostriatal caudate nucleus (*CN*) and putamen (*PT*), as well as with the accumbens nucleus (*AC*) of the limbic striatum. (**b**) Atrophy, pallor, and severe neuronal loss in the neostriatal CN and PT of a representative Huntington's disease (HD) patient with Vonsattel grade 4 of neostriatal atrophy. Note the markedly enlarged lateral ventricle (V), the atrophic internal capsule (*IC*), as well as the markedly concave medial outlines of the CN and *IC*. Although also atrophic and pale, the AC of the limbic striatum of the HD patient was less severely affected by neuronal loss than the neostriatal CN and PT. (**a**, **b**: gallocyanin staining, 400 μm gelatine sections). Abbreviations: *AC* accumbens nucleus, *CC* corpus callosum, *CL* claustrum, *CN* caudate nucleus, *IC* internal capsule, *PT* putamen, *V* lateral ventricle

mainly caused by the dramatic loss of neostriatal GABAergic medium spiny stellate projection neurons. These nerve cells represent approximately 95 % of all neostriatal neurons; coexpress substance P, enkephalin, and dynorphin; and constitute the main targets of striatal input and the main source of striatal output. As compared to these neostriatal GABAergic nerve cells, large cholinergic neurons which coexpress somatostatin and NPY are less severely affected in HD (Borrell-Pagès et al. 2006; Bruyn et al. 1979; De la Monte et al. 1988; Dom et al. 1976; Dunlap, 1927; Duyao et al. 1993; Estrada-Sanchez and Rebec 2013; Fennema-Notestine et al. 2004; Ferrante et al. 1987; Finkbeiner and Mitra 2008; Hedreen et al. 1991; Heinsen et al. 1994, 1996, 1999; Heinsen and Rüb 1997; Imarisio et al. 2008; Lange 1981; Lange et al. 1976; Li and Conforti 2013; Margolis and Ross 2003; Myers 2004; Myers et al. 1988; Nakano et al. 2000; Parent 1990; Schulte and Littleton 2011; Sotrel et al. 1991; Valera et al. 2005; Vogt and Vogt 1920, 1942; Vonsattel 2008; Vonsattel and DiFiglia 1998; Vonsattel et al. 1985; Walker 2007a, b).

2.3 Functional Neuroanatomy of the Striatum

According to its major intracerebral connections and functional aspects, the human striatum has been divided into "sensorimotor," "associative," and "limbic" territories (Fig. 2.9). The *sensorimotor territory of the striatum* comprises the dorsolateral rims of the precommissural and postcommissural putamen, as well as the lateral part of

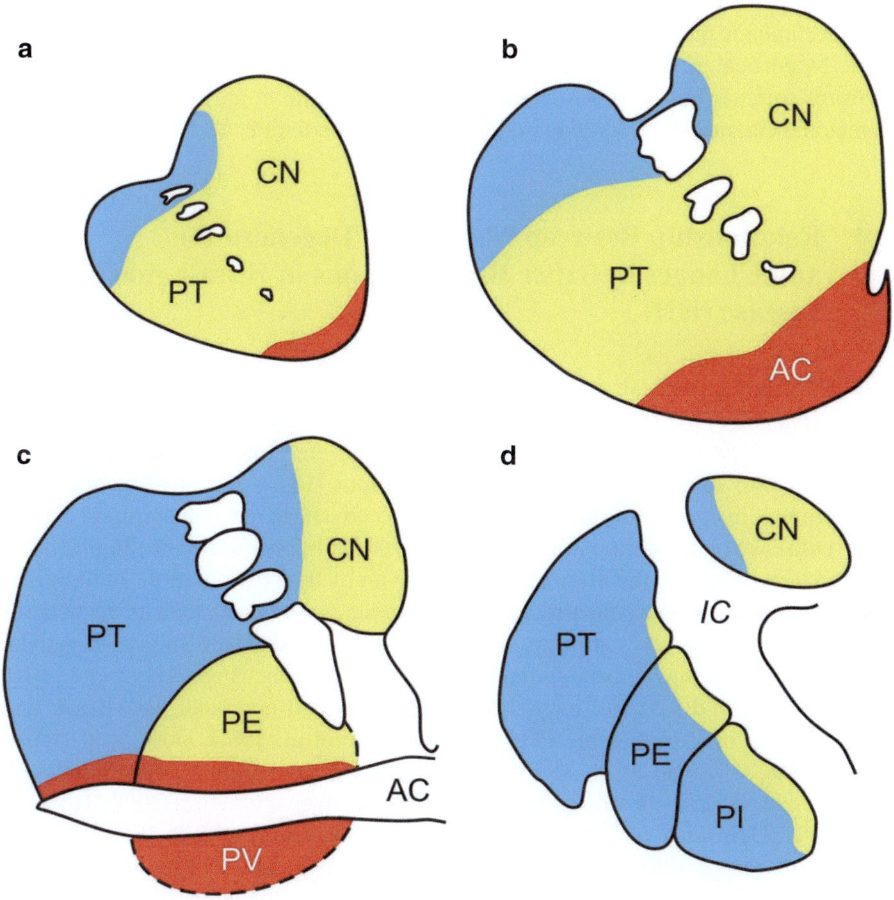

Fig. 2.9 The sensorimotor, associative, and limbic territories of the human striatum. Schematic frontal sections through different levels of the left human striatum showing its functionally defined "sensorimotor" (*blue*), "associative" (*yellow*), and "limbic" (*red*) territories. (**a**) Most rostral level of the caudate nucleus (*CN*) and putamen (*PT*), (**b**) level of the accumbens nucleus (*AC*), (**c**) level of the anterior commissure (*AC*) and ventral pallidum (*PV*), and (**d**) level of the external (*PE*) and internal pallidum (*PI*). For purposes of simplicity zones of overlap of the three striatal territories have been neglected (Modified according to Parent (1990), (Figure 1, page 255); with kind permission from Elsevier). Abbreviations: *AC* accumbens nucleus, *AC* anterior commissure, *CN* caudate nucleus, *IC* internal capsule, *PE* external pallidum, *PI* internal pallidum, *PT* putamen, *PV* ventral pallidum

the body of the caudate nucleus. It receives cortical afferents from the motor, premotor, supplementary, and cingulate cortical areas as well as from the somatosensory cortex. The *associative territory of the striatum*, however, occupies most of the head of the caudate nucleus, the medial aspects of the body and the tail of the caudate nucleus, the precommissural putamen, as well as the ventromedial parts of the postcommissural putamen. Cortical afferents to the associative territory of the striatum originate from associative areas of the cerebral cortex such as the prefrontal, temporal, posterior parietal, and preoccipital cortices, as well as from the oculomotor frontal and supplementary eye fields. The *limbic* (*or ventral*) *territory of the striatum* (i.e., accumbens nucleus, ventromedial part of the precommissural putamen) is targeted by cortical afferents from allo- and periallocortical areas (e.g., hippocampus, piriform cortex) (Fig. 2.9) (Groenewegen 2003; Nakano et al. 2000; Parent 1990; Parent and Hazrati 1995; Sadikot and Rymar 2009; Sadikot et al. 1992a, b).

2.4 Relationship Between Neostriatal Degeneration and Changes in Other Brain Regions in Huntington's Disease (HD)

The HD grade of striatal atrophy correlates with the extent of the atrophy of the *regions outside of the striatum*. In grades 1 and 2, non-striatal structures of the brain which may show *macroscopical visible atrophic changes* include the cerebral cortex and white matter, as well as the cerebellum (Figs. 2.1d–f, 2.6, and 2.7). Furthermore, in carriers of the IT15 gene with an advanced age, concomitant degenerative disease (e.g., brain pathologies related to Alzheimer's disease; Fig. 2.1d–f) may be incremental to the brain alterations driven by the HD mutation. In individuals with grades 3 or 4 striatal atrophy, *non-neostriatal structures* such as the neocortex, allocortex, thalamus, pallidum, subthalamic nucleus, substantia nigra, amygdala, or the cerebellum may appear slightly or markedly atrophic (Figs. 1.4, 2.1j–l, 2.10, 2.11, 3.1, 3.2, and 3.3) and may show mild or marked neuronal loss (Braak and Braak 1992a, b; Bruyn et al. 1979; De la Monte et al. 1988; Dom et al. 1976; Dunlap, 1927; Duyao et al. 1993; Estrada-Sanchez and Rebec 2013; Fennema-Notestine et al. 2004; Ferrante et al. 1987; Finkbeiner and Mitra 2008; Hedreen et al. 1991; Heinsen et al. 1992, 1994, 1996, 1999; Heinsen and Rüb 1997; Imarisio et al. 2008; Lange 1981; Lange and Aulich 1986; Lange et al. 1976; Margolis and Ross 2003; Myers 2004; Myers et al. 1988; Rüb et al. 2013a, 2014a, b; Selemon et al. 2004; Sotrel et al. 1991; Valera et al. 2005; Vogt and Vogt 1920, 1942; Vonsattel 2008; Vonsattel and DiFiglia 1998; Vonsattel et al. 1985; Walker 2007a, b).

Recently, the evaluation of a series of 523 HD brains that were all processed according to the same standardized neuropathological protocol (Hadzi et al. 2012) revealed a disproportional striatal atrophy compared to the relatively preserved non-striatal regions. In view of this varying extent of the atrophy of the striatum versus that of non-striatal hemispheric regions, two clusters were identified in HD brains (i.e., a *cortical cluster and a striatal cluster*). While the striatal atrophy was found

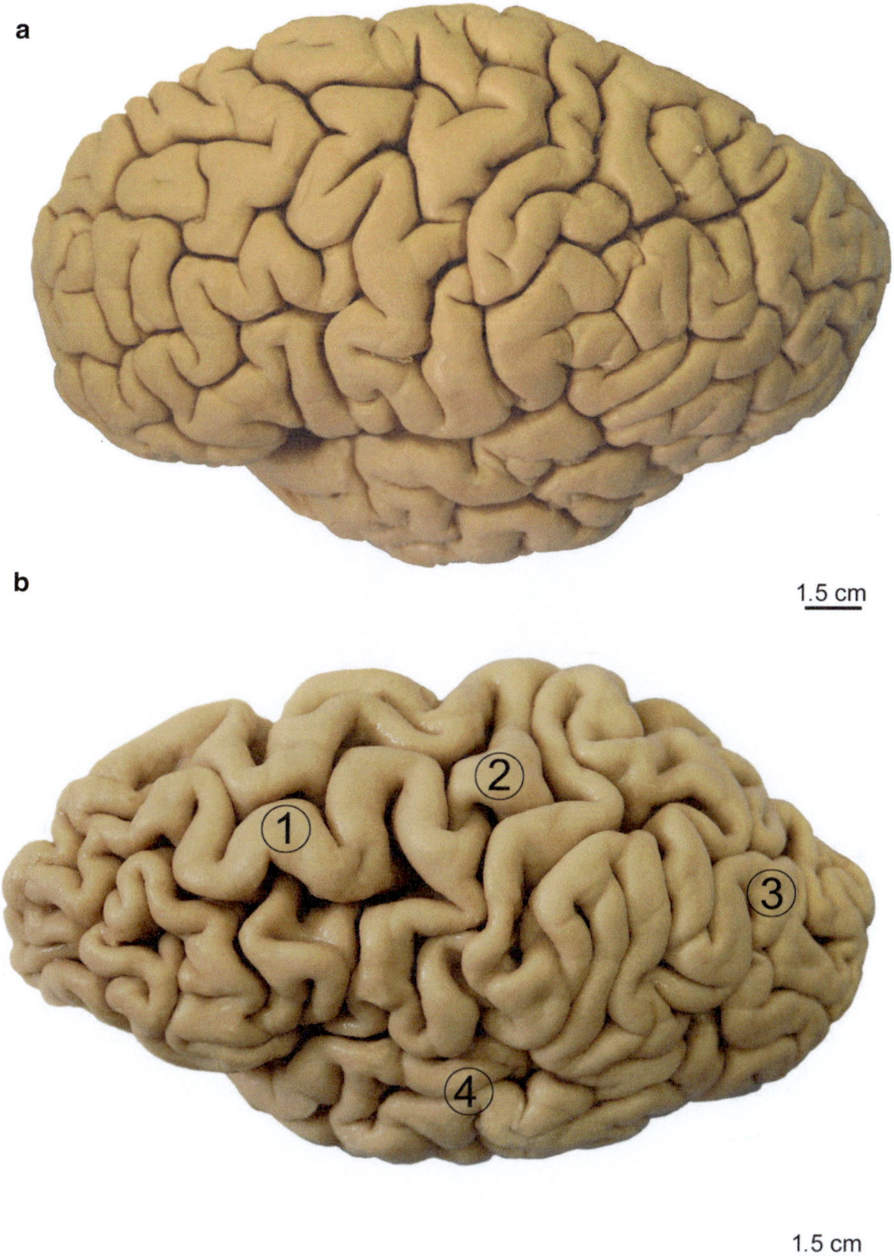

Fig. 2.10 Atrophy of the cerebral hemisphere in Huntington's disease (HD). (**a**) Dorsolateral aspect of the left cerebral hemisphere of a typical control individual without a medical history of neurological or psychiatric diseases. (**b**) Dorsolateral aspect of the left cerebral hemisphere of a clinically diagnosed and genetically confirmed male Huntington's disease patient (Vonsattel grade 2 of neostriatal atrophy). Note the marked atrophy of the frontal lobe (*1*), pericentral region (*2*), occipital lobe (*3*), and temporal lobe (*4*) (Reprinted from Rüb et al. (2009), (Figure 1, page 5); with kind permission from John Wiley and Sons)

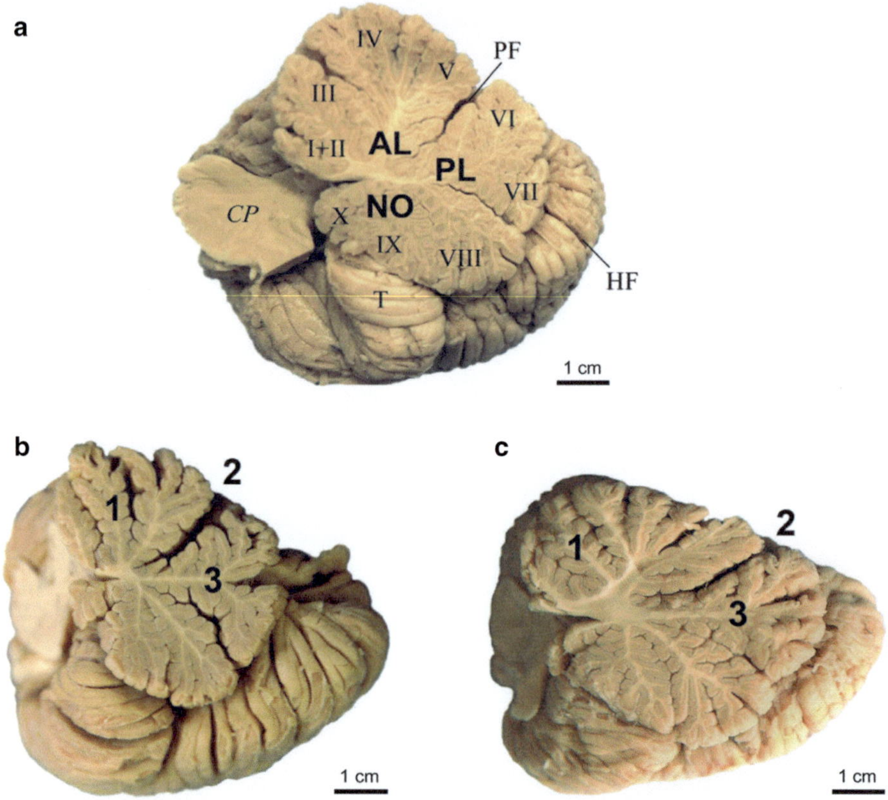

Fig. 2.11 Macroscopical changes of the cerebellum in Huntington's disease (HD). (**a**) Medial view of the right cerebellar hemisphere of a typical control individual. (**b**, **c**) The atrophic right cerebellar hemisphere of two representative Huntington's disease (HD) patients (the neostriatal atrophy in the brains of both HD patients corresponded to Vonsattel grade 3). Both atrophic cerebella showed an overall volume loss, a reduced arbor vitae, atrophic lobules of the anterior (*1*) and posterior lobes (*3*), as well as a widened primary fissure (*2*) (Reprinted from Rüb et al. (2013a), (Figure 1, page 169); with kind permission from John Wiley and Sons). Abbreviations: *AL* anterior lobe, *HD* Huntington's disease, *Lobules I, II, and III* lingula and central lobule, *Lobules IV and V* culmen, *Lobule VI* declive, *Lobule VII* tuber and folium, *Lobule VIII* pyramis, *Lobule IX* uvula, *Lobule X* nodule, *CP* cerebellar peduncles, *NO* nodule, *PL* posterior lobe, *HF* horizontal fissure, *PF* primary fissure, *T* cerebellar tonsil

to parallel to some extent that of the non-striatal hemispheric regions in the cortical cluster, the severity of striatal atrophy was disproportional compared to that of the non-striatal hemispheric regions in the striatal cluster. Although both HD brain clusters were associated with the length of the CAG repeats in the HD gene, the striatal cluster correlated more than the cortical cluster with the length of the CAG repeats, age at HD onset, and age at death. In contrast, the cortical brain cluster showed a stronger correlation than the striatal cluster with decreased weight of the HD brains. These distinct morphological phenotypes might result from differential regional pathophysiological mechanisms, differential regional vulnerability, and cellular susceptibility to the toxic effects driven by the HD mutation.

Since details on changes occurring outside of the striatum of HD brains are thoroughly described elsewhere (see Chaps. 3, 4, 5, 6, 7, and 11), only a brief survey of the alterations involving the *pallidum* is addressed here. The pallidum undergoes initial neurodegeneration during grade 2, which further progresses toward grades 3 and 4 and results in a volume loss of nearly 50 % in grade 4 brains (Figs. 1.4 and 2.1e, h, k). In individuals with grade 2, 3, and 4 of striatal neuropathology, the external segment of the pallidum is commonly more severely involved than its internal segment (Fig. 1.4). Microscopically, the pallidum is less abnormal than one would expect from the degree of its macroscopically visible atrophy. In brains with grade 3 and 4 of striatal atrophy, the neurons of the pallidum are smaller and more densely packed than normal suggesting widely preserved nerve cells, but shrinkage of the pallidal neuropil (Vonsattel 2008; Vonsattel and DiFiglia 1998; Vonsattel et al. 1985). Since Lange et al. reported that (1) the absolute number of pallidal neurons decreases up to 40 % during HD, (2) the neuronal density is up to 42 % higher than normal in the external pallidum and 27 % higher in the internal pallidum (Lange et al. 1976), atrophy of the pallidum in HD appears mainly due to loss of neuropil, encompassing striatal fiber connections, and fiber passages and to a lesser extent due to loss of neurons (Fig. 1.4) (Campbell et al. 1961; McCaughey 1961; Neustaedter 1933; Schroeder 1931; Spielmeyer 1926; Vonsattel et al. 1985). Reactive astrogliosis can be reliably detected in the pallidum of grade 3 brains, worsens in grade 4, and is usually confined to the external segment of the pallidum.

This chapter was written by Jean Paul Vonsattel and Udo Rüb.

Chapter 3
The Cerebral Cortex in Huntington's Disease (HD)

3.1 Establishment of Cortical Neuronal Loss as a Main Neuropathological Hallmark of Huntington's Disease (HD) and Its Clinical Relevance

As described in Chap. 2, the neuropathological hallmark of Huntington's disease (HD) is *progressive striatal degeneration*, which according to the Vonsattel grading system can be classified into five grades in ascending order of severity of striatal degeneration (Vonsattel 2008; Vonsattel and DiFiglia 1998; Vonsattel et al. 1985). While the affection of the striatum is acknowledged as neuropathological feature of HD since more than a hundred years, the affection of the cerebral cortex during the course of HD and its relevance for the clinical picture were a matter of debate for a long time (see Chaps. 1 and 2). However, a large variety of quantitative and qualitative postmortem studies repeatedly demonstrated *cortical neuronal loss* in HD patients, which is commonly most pronounced in the isocortical neuronal layers III, V, and VI (Brodmann 1909; Bruyn 1968; Bruyn et al. 1979; Estrada-Sanchez and Rebec 2013; Fennema-Notestine et al. 2004; Forno and Jose 1973; Hedreen et al. 1991; Heinsen et al. 1992, 1994, 1999; Heinsen and Rüb 1997; McCaughey 1961; Richardson 1990; Selemon et al. 2004; Sotrel et al. 1991; Vonsattel 2008; Vonsattel and DiFiglia 1998).

The findings of these postmortem studies eventually allowed to establish the affection of the *cerebral cortex as a main neuropathological feature of HD* (Figs. 3.1, 3.2, and 3.3), which is currently regarded as the main morphological correlate of psychiatric symptoms, neuropsychological deficits, cognitive decline, and neurophysiological abnormalities of HD patients and some at-risk gene carriers. According to current pathophysiological ideas, the affection of the cerebral cortex not only represents the morphological correlate of these psychiatric, neuropsychological, and neurophysiological manifestations in HD but may at least be jointly responsible

U. Rüb et al., *The Neuropathology of Huntington's Disease: Classical Findings, Recent Developments and Correlation to Functional Neuroanatomy*,
Advances in Anatomy, Embryology and Cell Biology, Vol. 217,
DOI 10.1007/978-3-319-19285-7_3

Fig. 3.1 Laminar neuronal loss and white matter changes in the temporal neocortex of Huntington's disease (HD) patients. (**a**) Thick stained frontal tissue sections through the occipitotemporal neocortical region (Brodmann's area 37 or area PH according to von Economo and Koskinas) of a representative control individual, (**b**) a typical Huntington's disease (HD) patient, and (**c**) a typical Alzheimer's disease (AD) patient (Brodmann 1909; von Economo and Koskinas 1925). (**d**) 12 μm-thick gallocyanin-stained paraffin section parallel to the thick section shown in (**b**). (**b**) Neuronal loss and pallor of pyramidal cell layers III and V (*black arrows*) and a distinct well-preserved and prominent layer IV (*arrowheads*) are pathognomonic not only for the temporal association fields but also for all isocortical association fields in the frontal, parietal, and occipital lobes of HD patients (Heinsen et al. 1994). In gallocyanin-stained sections, the cerebral white matter of HD patients commonly shows an intensive staining (i.e., chromophilia). The white matter chromophilia (*white arrows*) in HD could reflect ongoing attempts of remyelination (Bartzokis et al. 2007) and may be linked to the increased oligodendroglial density in the caudate nucleus of early- and low-grade HD patients (Gómez-Tortosa et al. 2001; Myers et al. 1991). The medullary layer abutting on layer VI and most likely comprising associative u-fibers (*asterisks*) is intact in HD patients and stains similarly like those of the control individuals. (**c**) Less conspicuous neuronal loss in layers III and VI and more pronounced neurodegeneration in layer V of the occipitotemporal neocortical region of the AD patient. (**d**) Whereas unconventionally thick brain tissue sections facilitate diagnosis of cortical neuronal loss in HD, laminar neurodegeneration is difficult to assess in conventional 5–20 μm-thick paraffin sections. Scale bar in D is also valid for **a**–**c**. (**a**–**c**, gallocyanin staining, 440 μm gelatine sections; **d**, gallocyanin staining, 12 μm paraffin section)

for the occurrence of *involuntary choreatic movements* (Sheperd 2013). Choreatic movements representing the primary involuntary movement abnormalities of HD patients evolve insidiously during the transition from the asymptomatic to the symptomatic phase of HD. They increase and become generalized, abrupt, and uncontrolled only in a second clinical HD stage and then ultimately plateau and wane during the later disease course of HD (Estrada-Sanchez and Rebec 2013; Margolis and Ross 2003; Vonsattel and DiFiglia 1998; Walker 2007a, b). The pathognomonic choreatic movements very frequently are preceded during the prediagnostic phase by a spectrum of psychiatric and/or neuropsychological HD symptoms. Subsequent to their onset, *psychiatric and neuropsychological HD symptoms* commonly progress concomitantly with the worsening of choreatic movements and finally culminate in severe psycho-organic syndromes or dementia (Estrada-Sanchez and Rebec 2013; Margolis and Ross 2003; Shoulson and Young 2011; Vonsattel and DiFiglia 1998; Walker 2007a, b). Owing to the very early onset of these progressive nonmotor HD symptoms, it is widely believed that the affection of the cerebral cortex may take place even before the striatum undergoes neurodegeneration (Estrada-Sanchez and Rebec 2013; Margolis and Ross 2003). Traditionally, the occurrence of chorea has been widely regarded as an immediate consequence of striatal degeneration (Estrada-Sanchez and Rebec 2013; Mink 1996; Walker 2007a, b), but already Cécile and Oskar Vogt challenged this opinion and assumed that the cortical pathology may play a substantial pathophysiological role in the evolution of chorea (Vogt and Vogt 1920). The traditional understanding of the nature and function of the striatum has been recently unmasked as a circular argument derived from clinical observations (Walker 2007a). Since comprehensive analysis revealed that lesions of the caudate nucleus, putamen, or pallidum result in choreatic movements only in

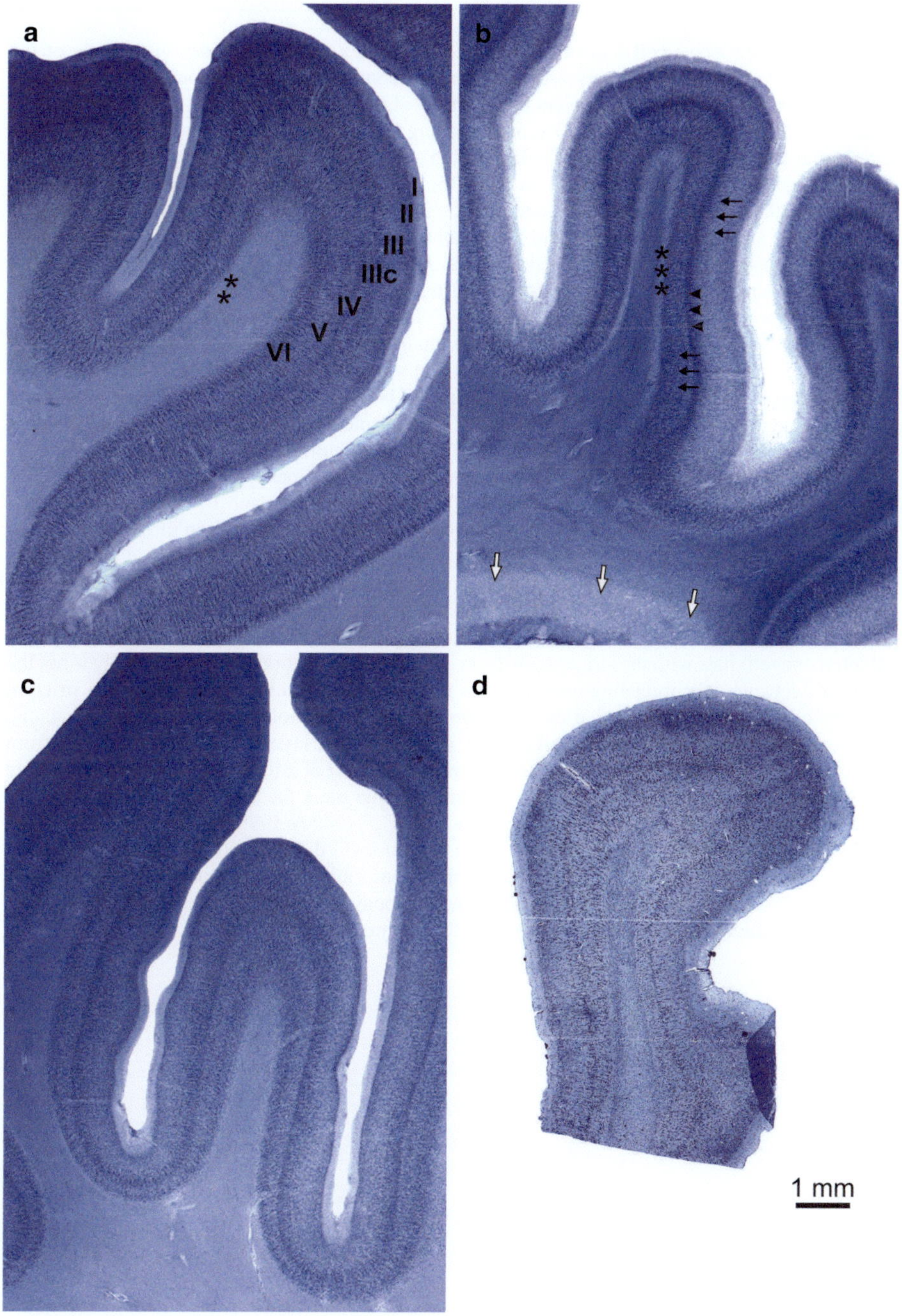
a
I
II
III
IIIc
IV
V
VI
**
b
***
c
d
1 mm

Fig. 3.2 Laminar nerve cell loss in the temporal neocortex and the primary visual area of the occipital lobe in Huntington's disease (HD). (**a**) Unconventional thick frontal tissue sections through the infragranular cortical layers in the crown of the posterior parts of the middle temporal gyrus covered by Brodmann's area 37 (corresponding to area PH of von Economo and Koskinas) (Brodmann 1909; von Economo and Koskinas 1925) of a typical control case and (**b**) a representative Huntington's disease (HD) patient. (**a**) Note the striking columnar arrangement of the cortical neurons and their separation by unstained radially coursing fiber bundles. (**b**) Owing to neuropil shrinkage, the surviving layer IV, V, and VI neurons in Brodmann's area 37 of the HD patient are more closely spaced. Although the total neuronal number is reduced in these cortical nerve cell layers, the apparently higher neuronal density, particularly in layer IV, suggests unchanged neuronal numbers thereby masking true neuronal loss. (**c**) Thick frontal tissue section through the caudal pole of the occipital lobe of a typical control individual at the border between Brodmann's primary visual area 17 and the adjacent parastriate Brodmann's area 18 (*dashed vertical line*). The "calcarine cortex type" of Brodmann's primary visual area 17 is characterized by: (1) a striking reduction of the outer and inner pyramidal layers III and V, (2) conspicuous granular layers II and IV with a particularly high neuronal density, (3) a strongly basophilic multiform layer VI, and (4) the subdivision of the inner granular layer into layers IVa–c. Note the prominent pale, cell-poor, and myelin-rich Gennari stripe in the inner granular layer IVb (*asterisks*) (Brodmann 1909; Garey 1990; Rüb et al. 2014b; Zilles 1990). (**d**) High-power photograph of a thick tissue section through the degenerated primary visual area 17 of a typical HD patient: (1) severe nerve cell loss and pallor of the outer pyramidal layer III, (2) markedly reduced nerve cell density and pallor of the inner granular layers IVa and IVc, (3) narrowed multiform layer VI reflecting nerve cell loss. Asterisks indicate Gennari stripe (Rüb et al. 2014b). Since Brodmann's primary visual area 17 has no direct anatomical interconnectivities with the striatum, its layer-specific degeneration in HD most likely is not the result of a retrogradely spreading neuronal loss following the initial degeneration in the striatum but may be a primary pathological event. *White box* in (**d**) symbolizes the six-layered thalamic lateral geniculate body (*m* magnocellular layer, *p* parvocellular layer, *k* koniocellular layer) from where the granular layers II and IV and the pyramidal layer III of Brodmann's primary visual area 17 receive visual projections via the optic radiation. *Arrows* mark the laminar targets of these subcortical projections in Brodmann's primary visual area 17 (Garey 1990; Rüb et al. 2014b; Zilles 1990). (**a–d**: gallocyanin staining, 440 μm gelatine sections). (**d**: Modified according to Rüb et al. (2014b), (Figure 1, page 169); with kind permission from John Wiley and Sons). Abbreviations: *m* magnocellular layer of the thalamic lateral geniculate body, *p* parvocellular layer of the thalamic lateral geniculate body, *k* koniocellular layer of the thalamic lateral geniculate body

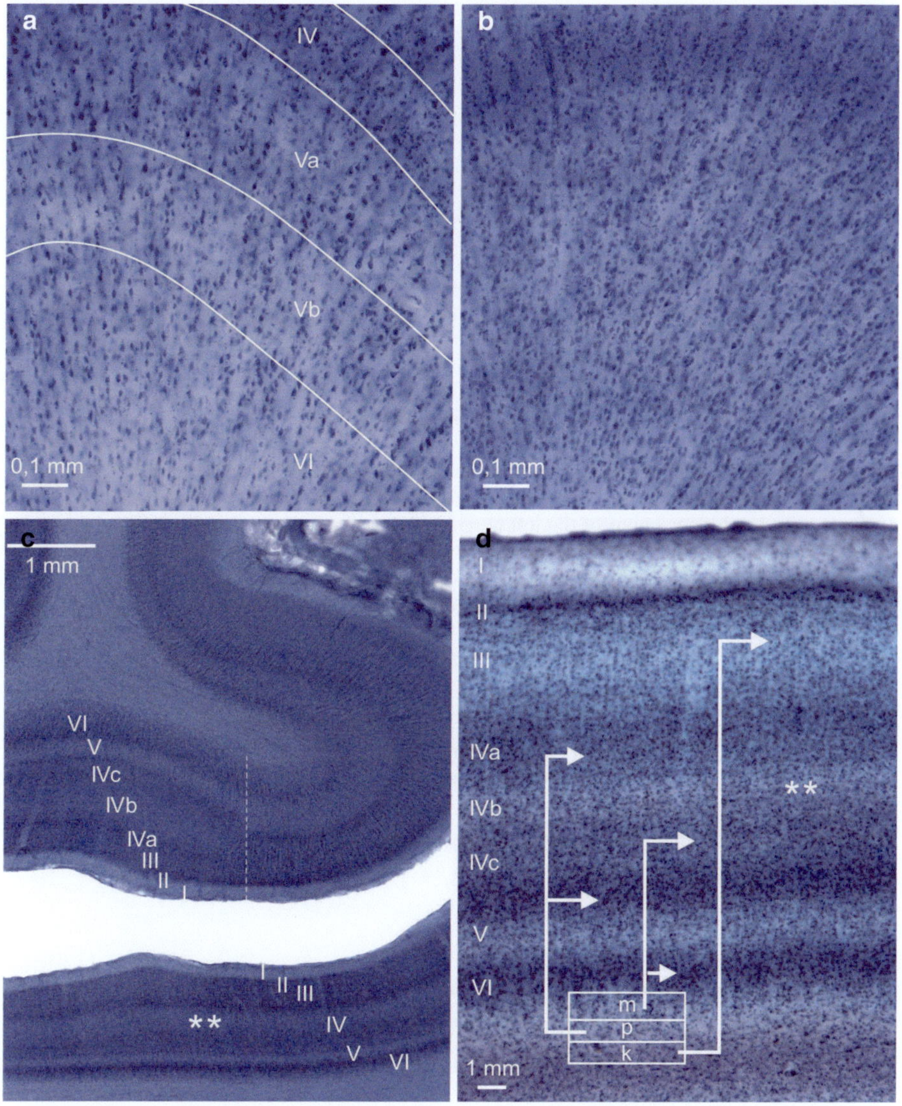
a
IV
Va
Vb
VI
0,1 mm
b
0,1 mm
c
1 mm
VI
V
IVc
IVb
IVa
III
II
I
I
II
III
**
IV
V
VI
d
I
II
III
IVa
IVb
IVc
V
VI
**
m
p
k
1 mm

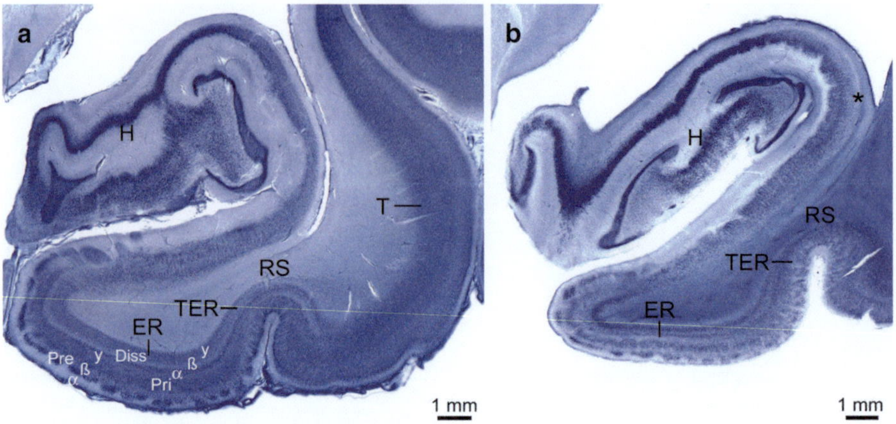

Fig. 3.3 Laminar neuronal loss in the allocortical transentorhinal and entorhinal regions in Huntington's disease (HD). (**a**) Thick gallocyanin-stained frontal section through the parahippocampal gyrus and hippocampus of the left temporal lobe of a typical control individual and (**b**) a representative Huntington's disease patient. (**b**) Moderate neuronal loss in the subiculum of the HD patient (*asterisk*) and nearly complete degeneration of the pre-β and pri-γ layers in the entorhinal (*ER*) (Brodmann's area 28) and transentorhinal (*TER*) (Brodmann's area 35) regions (Braak and Braak 1992a, b; Brodmann 1909; Heinsen et al. 1992). (**a**, **b**: gallocyanin staining, 440 μm gelatine sections). Abbreviations: *Diss* lamina dissecans, *ER* entorhinal cortex, *H* hippocampus, *Pre* principal external layer, *Pri* principal internal layer, *RS* rhinal sulcus, *T* temporal neocortex, *TER* transentorhinal cortex

the minority of affected individuals (Bhatia and Marsden 1994), the manifestation of chorea is currently supposed to result from dysfunctional corticostriatal projection neurons and/or their impaired intra-telencephalic input in HD (Estrada-Sanchez and Rebec 2013; Sheperd 2013).

3.2 Confirmation and Extension of Very Early Descriptions of Cortical Neurodegeneration in Huntington's Disease (HD) by Modern Quantitative Approaches

Initial conjectures, qualitative assessments, and the first quantitative attempts around the turn of the century already suggested degeneration and nerve cell loss in the cerebral cortex of HD patients (see Chap. 1). However, it was only in the 1980s when the *first quantitative postmortem investigations* of the brains of HD patients were performed by means of more sophisticated methods. According to our and others' experience, the true nerve cell loss in a given brain region can be unequivocally detected upon qualitative light-microscopical investigation only, if its extent exceeds a threshold of approximately 60 %. To investigate early neurodegenerative events on a solid base and to obtain insight into neurodegenerative events that are less severe than those in the striatum, quantitative morphometric methods were applied to assess volumes, cross-sectional areas, and widths, as well as cell densities in the cerebral cortex and associated white matter in the brains of HD patients.

The first pioneering and groundbreaking volumetric studies of serial tissue sections through the brains of HD patients yielded *atrophy and volume loss of distinct cortical areas and regions*. Cortical volume loss was 30 % in the peristriate and parastriate visual areas in the occipital lobe (i.e., Brodmann's areas 18 and 19), 17 % in the primary visual striate area covering the calcarine sulcus in the occipital lobe (i.e., Brodmann's area 17), 19 % in the primary somatosensory cortex (i.e., Brodmann's areas 1, 2, and 3), 12–21 % in the frontal cortex, 15 % in the anterior cingulate cortex (i.e., Brodmann's area 24) and 25 % in the posterior cingulate cortex (i.e., Brodmann's area 23), 12 % in the primary auditory cortex (i.e., Brodmann's areas 41 and 42), 12 % in the insula, 10 % in the temporal neocortex, and 10 % in the parahippocampal gyrus with the transentorhinal and entorhinal regions (Brodmann 1909; Lange 1981; Lange and Aulich 1986). In addition to detailed findings in the cerebral cortex, these accurate morphometric studies also showed degenerative signs in subcortical brain regions of HD patients, which stimulated further quantitative HD studies that directly or indirectly confirmed the initial observations in cortical and subcortical regions (Fig. 1.4). Taken together, the findings of these accurate morphometric studies contrasted with the *traditional reductionistic viewpoint of HD* as a degenerative disease predominantly affecting the striatum and pointed to a widespread but area-specific vulnerability of the cerebral cortex in HD. Representing milestones of neuropathological HD research, these groundbreaking morphometric studies have definitely changed the traditional unilateral pathoanatomical and pathophysiological concepts of HD, which solely took the affection of the striatum into consideration. They contributed substantially to the paradigm shift in neuropathological HD research, which ultimately paved the way for the breakthrough of the concept of *HD as a multisystem neurodegenerative disease* (Heinsen et al. 1994; Lange and Aulich 1986; Rüb et al. 2013a; Vonsattel 2008).

Estimations of brain weights disclosed a reduction of the total brain weight by 30 % in the HD individuals studied, and evaluation of brain *cross-sectional areas* found a mean reduction of the cross-sectional area of the cerebral cortex by 21–25 % and of the underlying cerebral white matter by 29–34 %, both of which correlated negatively with higher grades of HD severity (De la Monte et al. 1988). Ten years later the results on the decline of brain weight and reduction of the cross-sectional areas of the cerebral cortex and white matter of HD brains were confirmed and extended by the findings of a reduced *length and thickness of the cerebral cortex* or some of its areas in HD. A comparison of these morphometric postmortem results suggested that loss of cerebral white matter in the brains of HD patients is more severe than the reduction of cerebral gray matter (Mann et al. 1993). In 1991, Hedreen et al. reported a statistically significant decrease of the total cortical thickness and widths of cortical layers III, V, and VI of the dorsal bank of the superior frontal sulcus in HD patients which belongs to Brodmann's cortical area 10 (Brodmann 1909; Hedreen et al. 1991).

Further analysis of the brain volumes pointed to a widespread and global disease process in HD patients and demonstrated severe brain atrophy resulting from volume reductions in both cortical and subcortical gray matters (e.g., striatum). Although the cortex of the medial temporal lobe seemed to be more or less spared, cortical atrophy was relatively uniformly distributed over the four cerebral lobes (Fig. 2.10). Striatal atrophy correlated with the severity of cortical atrophy and loss of frontal white matter with both cortical and striatal atrophies (Halliday et al. 1998).

Reconsiderations of the theoretical background of the quantitative morphometric studies that were widely appreciated and applied in basic neurological and psychiatric research at that time led to the insight that estimations of volumes, widths, or cross-sectional areas of distinct brain regions may suffer from bias due to complex agonal, postmortem, and technical factors (e.g., brain tissue shrinkage). These factors can influence the morphometric parameters in an unpredictable manner; they may reflect atrophy of nerve cells, their processes, and/or other components of the brain neuropil in a false-positive or false-negative way. Since they cannot be effectively compensated by strict protocols of brain fixation, brain embedding, and staining procedures (Coggeshall and Lekan 1996; Heinsen and Heinsen 1991; Heinsen et al. 1992, 1994, 1996, 1999; Oorschot 1994; Schmitz and Hof 2005; West and Gundersen 1990), analyses without correction of the bias do not necessarily provide data that reliably substantiate true death of nerve or glial cells. As in other fields of neuroscience, estimations of volumes, widths, or cross-sectional areas of distinct brain regions were therefore complemented and replaced by morphometric estimations of neuronal densities in neuropathological HD research. These quantitative estimations of neuronal densities, however, have been regarded only for a limited time span as morphometric parameters that can furnish additional and better quantitative arguments in the assessment and diagnosis of brain atrophy and/or neurodegeneration (Coggeshall and Lekan 1996; Heinsen and Heinsen 1991; Heinsen et al. 1992, 1994, 1996, 1999; Oorschot 1994; Schmitz and Hof 2005; West and Gundersen 1990).

One of the earliest morphometric postmortem studies which was based on the estimation of nerve cell densities described a significant reduction of the density of pyramidal projection neurons immunoreactive for nonphosphorylated neurofilament proteins (SMI-32) and reported a normal density of cortical nerve cells immunoreactive for the neuropeptide Y (NPY) in Brodmann's areas 8, 9, 11, and 22 of HD patients (Brodmann 1909; Cudkowicz and Kowall 1990). Subsequently, the reduction of the density of SMI-32 immunoreactive nerve cells was also proven in the angular gyrus of HD patients located at the junction of the temporal, parietal, and occipital lobes (i.e., Brodmann's area 39) (Brodmann 1909; MacDonald et al. 1997).

Based on stereological countings of various cortical cell types (i.e., large pyramidal neurons, small neurons, astrocytes, and oligodendroglial and microglial cells), recording of the thickness of each neuronal layer of the cerebral cortex, and estimations of nerve cell densities, a detailed postmortem study performed on Brodmann's prefrontal areas 8 and 9 of HD patients extended the findings of Cudkowicz and Kowall and demonstrated a selective and progressive loss of a subset of the large pyramidal neurons in cortical layers III, V, and VI, which was accompanied by a decrease in the thickness of these cortical neuronal layers. As in previous and subsequent HD studies, degeneration of cortical pyramidal neurons was interpreted either as a primary event in the pathological process of HD or as the result of a retrograde process induced by a primary degeneration in the striatum (Sotrel et al. 1991).

An additional quantitative postmortem HD study employed three-dimensional cell counting methods for the assessment of laminar cell densities, cell numbers in select cortical areas, and laminar widths. This study discovered an overall reduction of the cortical thickness of Brodmann's areas 9 (26 %) and 46 (23 %); a reduction of the density of projection neurons in layers III (16 %), V (31 %), and VI (37 %); and a reduced width of these layers (III–25 %, V–34 %, and VI–46 %) in Brodmann's area 9. In contrast, reduction of the width of cortical layers II (18 %) and VI (35 %) was not accompanied by neuronal loss. In addition, the density of glial cells was significantly increased in layer VI (68 %) of Brodmann's area 9 and in layer V (75 %) of Brodmann's area 46. In view of the disparity in prefrontal cortical neurodegeneration, the authors suggested that (1) the prefrontal neurodegeneration underscores the dissociation of cortical neuronal loss from striatal retrograde degeneration, (2) the affection of the cerebral cortex in HD may not result from retrograde degeneration induced by the striatal pathology, (3) retrograde corticostriatal degeneration is not the driving force behind the cortical pathology in HD, and (4) the pathological changes in the cerebral cortex are produced by a cascade of downstream neurodegeneration in the cortico-striato-thalamocortical loops and eventually triggered by neurodegeneration of thalamic nuclei (Brodmann 1909; Selemon et al. 2004).

These quantitative postmortem HD studies confirmed and extended the early findings about the state of the cerebral cortex in HD patients from the 1990s, provided additional evidence for an overall neuronal loss in the cerebral cortex and the differential vulnerability of its distinct areas, and supported the arising opinion that degeneration of the cerebral neo- and allocortex may be among the neuropathological hallmarks of HD and of substantial relevance for the clinical picture of HD.

3.3 Application of Advanced, Unbiased Stereological Methods for the Assessment of Nerve Cell Loss

The field of quantitative brain research methods, however, progressed further. Approximately 20 years ago, additional efforts culminated in the design and implementation of sophisticated stereological methods that facilitate unbiased estimations of total cell numbers, which are not confounded by agonal, postmortem, or technical factors and therefore reflect their real extent in the brain region studied (Coggeshall and Lekan 1996; Heinsen and Heinsen 1991; Heinsen et al. 1992, 1994, 1996, 1999; Oorschot 1994; Schmitz and Hof 2005; West and Gundersen 1990).

One of the first *unbiased estimations of absolute nerve cell and glial cell numbers in the cerebral cortex* of HD patients was performed 20 years ago on unconventional thick serial gallocyanin-stained frontal hemispheric sections and involved Cavalieri's principle for volume and the optical disector for cell density estimations (Coggeshall and Lekan 1996; Heinsen and Heinsen 1991, 1992, 1994, 1996, 1999; Oorschot 1994; Schmitz and Hof 2005; West and Gundersen 1990). Unfortunately, the convincing results of this carefully performed and sophisticated postmortem study were widely neglected for nearly 10 years, but to date this study can be regarded as an important milestone of neuropathological HD research. Along with a conspicuous atrophy of the striatum and a tremendous mean decrease of small striatal neurons of about 88 % in the HD patients ($9.72 \times 10^6 \pm 3.64 \times 10^6$) in comparison to the control individuals ($82 \times 10^6 \pm 15.8 \times 10^6$), this stereological study disclosed a mean reduction of the absolute cortical neuronal numbers in the HD patients studied by approximately 33 % (HD, $3.99 \times 10^9 \pm 218 \times 10^6$; control individuals, $5.97 \times 10^9 \pm 320 \times 10^6$ nerve cells) and an overall volume loss of the cerebral cortex of nearly 22 % (HD, 108 ± 12 cm^3; control individuals, 138 ± 8 cm^3). These results (Figs. 3.1 and 3.2) confirmed the findings of initial volumetric HD studies (Lange 1981; Lange and Aulich 1986). Upon qualitative pathoanatomical inspection of the cortical ribbon, a layer-specific pallor and neuronal loss were demonstrated in the primary sensory areas (i.e., layers III, IV, and VI of Brodmann's areas 3, 1, 2, primary somatosensory cortex; layers III, IV, and VI of Brodmann's area 17, primary visual cortex; layers III, IV, and VI of Brodmann's area 41, primary auditory cortex), as well as in the isocortical association fields. In the isocortical association fields of the temporal, frontal, and parietal lobes, the spared layer IV granule cells formed a conspicuous band between the degenerated layers III and V (Figs. 3.1 and 3.2) (Brodmann 1909; Heinsen et al. 1994). Using unbiased, time-consuming, and labor-intensive modern stereological methods, this postmortem investigation of the cerebral cortex in HD not only confirmed convincingly the early impressions and quantitative findings of an overall reduction of the nerve cell number in the cerebral cortex of HD patients (Cudkowicz and Kowall 1990; Sotrel et al. 1991) but also provided evidence for the *area- and layer-specific vulnerability of the cerebral cortex* suggested by former morphometric HD studies (De la Monte et al. 1988; Hedreen et al. 1991; Lange 1981; Lange and Aulich 1986; Mann et al. 1993; Sotrel et al. 1991).

Since, from the current viewpoint of morphometric brain research, most of the early quantitative cortical HD data can only be taken as indications of a real neuronal loss, the unbiased morphometric study of Heinsen et al. was the first that reliably showed that cortical neuronal loss is a decisive and intrinsic morphological feature of HD and is not only simulated or biased by cortical volume loss and shrinkage of the cortical neuropil and/or the cortical nerve cells (Heinsen et al. 1994). In addition, this important stereological study also strongly pointed to the area- and lamina-specific cortical vulnerability as an important pathophysiological factor in the pathogenesis of HD and therefore represents an indispensible milestone on the way to the final establishment of the affection of the cerebral cortex as a main neuropathological hallmark of HD and its relevance for a large spectrum of HD disease symptoms. Damage to specific areas of the cerebral cortex not only can serve as an appropriate explanation for the large spectrum of *psychiatric and neuropsychological disease symptoms and the cognitive decline of HD patients* but also can account for the abnormalities present upon neurophysiological investigations (i.e., *abnormalities of the somatosensory and visual evoked potentials*, SEP and VEP). Since abnormalities of the SEP and VEP have been reported in the early stages of HD, the affection of the primary somatosensory cortex covering Brodmann's areas 1, 2, and 3 and of the primary visual cortex in Brodmann's area 17 could be interpreted as terminal stages of an early-onset cortical neuronal degeneration in HD (Heinsen et al. 1994; Rüb et al. 2014b).

Recently an *unbiased estimation of the absolute nerve and glial cell numbers* was also performed on thick gallocyanin-stained frontoparallel tissue sections through the primary visual striate area in the calcarine sulcus of the occipital lobe (i.e., *Brodmann's area 17*) of HD patients (Brodmann 1909; Rüb et al. 2014b). Estimation of the absolute number of nerve and glial cells in Brodmann's area 17 was based on the application of Cavalieri's principle for volume and the optical disector for nerve and glial cell density estimations (Rüb et al. 2014b). This quantitative morphometric analysis showed a significant reduction of the estimated absolute nerve cell number in the primary visual cortex of the HD patients (71.044.037 ± 12.740.515 nerve cells) of 32 % in comparison to the control individuals (104.075.067 ± 9.424.491 nerve cells) and a significant reduction of the volume of Brodmann's area 17 of the HD patients by 21 %. Additional qualitative *pathoanatomical analysis* disclosed that nerve cell loss in Brodmann's area 17 was most prominent in the outer pyramidal layer III, the inner granular layers IVa and IVc, as well as in the multiform layer VI (Fig. 3.2c, d) (Brodmann 1909; Rüb et al. 2014b). The degree of nerve cell loss in the occipital striate area is in excellent agreement with the overall neuronal loss of 33 % previously estimated by Heinsen et al. in the cerebral cortex of HD patients (Heinsen et al. 1994) and underlines the reliability of the application of advanced and unbiased quantitative investigations in human brain research. The estimated volume reduction of 21 %, in addition, was in a good agreement with the volume loss of 17 % in Brodmann's area 17 assessed in HD brains of the C. and O. Vogt brain collection (Düsseldorf, Germany) by Lange and Aulich (Lange 1981; Lange and Aulich 1986). The estimated reduction of the absolute neuronal number in the primary visual cortex showed that the volume loss or atrophy of the striate cortex in HD patients is not only due to shrinkage of the neuropil and/or the nerve cells but also reflects a definite and intrinsic premature death of neurons (Rüb et al. 2014b).

Since direct anatomical connections between the primary visual striate area and the neostriatum do not exist, the neuronal degeneration in the primary visual area of HD patients might occur independently from the well-known striatal degeneration and could be interpreted as an early-onset and primary degenerative event rather than a secondary process subsequent to neurodegeneration of the striatum (Heinsen et al. 1994; Rüb et al. 2014b). Considering the anatomical interconnectivities of the predominantly affected neuronal layers of the primary visual area 17, it is conceivable that the disease process of HD also spreads to and has adverse effects for interconnected cortical and subcortical regions of the visual system. Such a retrograde or anterograde spread to other regions of the visual system is compatible with the hypothesis that HD represents a *human neurodegenerative disease with a multifocal or polytopic onset* (Heinsen et al. 1994; Lange and Aulich 1986; Rüb et al. 2013a, 2014b; Vonsattel 2008).

The primary visual cortex in the calcarine sulcus via the optic radiation receives the majority of incoming visual information flow from the retina and therefore represents the major neocortical port of entry of the human central visual system. Based on the current knowledge in the field of the functional neuroanatomy of the human central visual system, the pathological changes in Brodmann's primary visual area 17 most likely contribute substantially to the pathogenesis of the *basic and complex visual dysfunctions* observed in HD patients. Since the primary visual striate cortex has also been identified to be involved in the generation of the early components of the *visual evoked potentials (VEP)*, damage to this primary sensory field of the human brain may play an additional important pathophysiological role for the development of the typical VEP alterations in patients suffering from HD (Rüb et al. 2014b).

A combined postmortem determination of volumes and neuronal densities was also applied to obtain an unbiased estimation of the total numbers of projection and interneurons, as well as of the absolute number of SMI-32 immunopositive pyramidal cells in three *motor cortical areas* of HD patients. This stereological investigation of the motor cortices of HD patients (1) for the first time disclosed morphological alterations and loss of the SMI-32 immunopositive long-projecting pyramidal nerve cells in the primary motor cortex of HD patients and (2) implicated the selective degeneration in the corticofugal pathways in the pathogenesis of involuntary choreatic movements (MacDonald and Halliday 2002).

These unbiased quantitative findings in the *primary motor cortex* of HD patients were recently confirmed by a study which also employed modern stereological methods (i.e., Cavalieri's principle for volume estimation and the optical disector for determination of nerve cell densities) for the assessment of a real neuronal loss in the primary motor cortex (i.e., Brodmann's area 4) and the anterior cingulate cortex (i.e., Brodmann's area 24) of HD patients. The HD patients studied showed (1) a significant reduction of the total number of pyramidal cells of 24 % and of the subpopulation of SMI-32 immunoreactive pyramidal cells of 27 % in Brodmann's area 4 and (2) an overall loss of pyramidal cells of 36 % and of SMI-32 immunopositive pyramidal cells of 34 % in Brodmann's area 24. Neuronal loss in the primary motor cortex was associated with pronounced motor dysfunction of the HD patients studied and neuronal loss in the anterior cingulate cortex with their major mood symptomatology. Accordingly, these findings indicated that the heterogeneity in clinical symptomatology of HD may be related to a varying extent of cell loss in the corresponding functional regions of the cerebral cortex (Brodmann 1909; Thu et al. 2010).

A highly characteristic and predictable pattern of layer-specific neuronal loss has been demonstrated in qualitative and unbiased stereological postmortem studies of the *allocortical transentorhinal and entorhinal regions* of HD patients (Braak and Braak 1992a, b; Heinsen et al. 1992). This pattern included a dramatic layer-specific neuronal loss in the pre-ß and pri-γ layers (Fig. 3.3) and may provide an explanation for the volume loss in the parahippocampal gyrus of HD patients reported by Lange and Aulich (Lange and Aulich 1986). According to our experience, the neuronal loss in the *transentorhinal and entorhinal pre-ß and pri-γ layers* represents an outstanding, interindividually consistent, and reliable pathognomonic degenerative feature of the brains of HD patients (Heinsen et al. 1992). Owing to its tremendous dimensions, this impressive and unequivocal layer-specific allocortical neurodegeneration is always immediately detectable light-microscopically even at low magnification and therefore does not require additional confirmation by application of sophisticated stereological methods. In our experience detection of this dramatic layer-specific pre-ß and pri-γ neuronal loss alone is suggestive and pathbreaking for the neuropathological diagnosis of HD (Heinsen et al. 1992).

In contrast to HD, the nerve cells of the superficial transentorhinal and entorhinal pre-alpha cells are of particular vulnerability in *Alzheimer's disease (AD)*. They are widely believed to be the first cortical nerve cells which exhibit the disease-related tau cytoskeletal pathology, and in the advanced neuropathological Braak and Braak AD stages V and VI leave behind so-called ghost tangles, which indicate the original position of healthy pre-alpha nerve cells (Braak and Braak 1991, 1992a, b; Braak et al. 2006). Since in HD the entorhinal and transentorhinal pre-ß and pri-γ layers preferentially undergo neurodegeneration (Fig. 3.3), the laminar differences in neurodegeneration between HD and AD can serve as explanations for the different disease-specific *neuropsychological deficits of HD and AD patients* (Brandt et al. 1988; Butters et al. 1979; Heinsen et al. 1992, 1994; Pillon et al. 1993; Salmon et al. 1989).

Postmortem studies applying Golgi silver impregnations (Sotrel et al. 1993), immunohistochemical, neurochemical, and biochemical studies (Arzberger et al. 1997; DiProspero et al. 2004; Dunlop et al. 1992; Hassel et al. 2008; Modregger et al. 2002; Smith et al. 2007) provided additional evidence for accompanying tissue changes and further arguments for an HD-induced cortical degeneration.

3.4 Pathomechanisms of the Involvement of the Cerebral Cortex in Huntington's Disease (HD)

The results of qualitative and quantitative postmortem studies and neuroradiological in vivo studies from the last three decades repeatedly confirmed that the *cerebral cortex* is among the main targets of the degenerative process and sustains multiple degenerative alterations in HD (see Chap. 1). This involvement of the cerebral cortex actively participates in the widespread extension of the underlying disease process and accounts for a large spectrum of HD disease symptoms (i.e., motor symptoms, psychiatric symptoms, cognitive decline, neuropsychological deficits, neurophysiological abnormalities). *Layer-specific neuronal loss* has been discovered in the following neo- and allocortical regions of HD patients: (1) frontal cortex (outer pyramidal layer III, inner pyramidal layer V, multiform layer VI); (2) primary somatosensory cortex (outer pyramidal layer III, inner granular layer IV, multiform layer VI); (3) frontal, temporal, and parietal association cortices (outer pyramidal layer III, inner pyramidal layer V); (4) primary visual area 17 (outer pyramidal layer III, inner granular layers IVa and IVc, multiform layer VI); (5) primary auditory cortex (outer pyramidal layer III, inner granular layer IV, multiform layer VI); and (6) allocortical transentorhinal and entorhinal regions (pre-β and pri-γ layers) (Figs. 3.1, 3.2, and 3.3) (Braak and Braak 1992a, b; Brodmann 1909; Heinsen et al. 1992, 1994; MacDonald and Halliday 2002; Rüb et al. 2014b; Thu et al. 2010).

HD is traditionally regarded as a neurodegenerative disease in which the caudate nucleus is the primary target of the disease process from which this process subsequently spreads to other vulnerable brain sites by means of anterograde and/or retrograde mechanisms (see Chaps. 1 and 2). According to the currently favored view, this spread does not occur randomly but follows distinct anatomical pathways and thus affects anatomically interconnected brain sites in predictable chronological and topographical sequences. On the one hand, neurodegeneration of the cerebral cortex and layer-specific neuronal loss in a subset of its areas could reflect the long-lasting secondary effect of a retrogradely or anterogradely spreading neurodegeneration, which is induced by the initial damage to the striatum and/or extrastriatal subcortical regions and conveyed to interconnected cortical areas via brain fiber tracts. On the other hand, cortical degeneration may even present a primary event in the pathological process of HD, which secondarily triggers degeneration in interconnected cortical areas and/or the neostriatum via an anterograde expansion (Heinsen et al. 1994; Lange and Aulich 1986; Rüb et al. 2013a, 2014b). Although meanwhile a well-established neuropathological feature of HD, there are still many open questions related to the pathomechanisms of the affection of the cerebral cortex. Therefore, future studies should clarify (1) when, during the progressive striatal neurodegeneration, the neo- and allocortex become involved in HD, (2) whether the involvement of the cerebral cortex is a primary or secondary event in the cascades of the pathological process of HD, (3) whether the neuronal loss in select layers and areas of the neo- and allocortex is correlated with the laminar cortical neuronal population with efferents to or afferents from affected subcortical regions (i.e., neostriatum, thalamus), and (4) whether primary centers of degeneration actually exist in the cerebral cortex or some of its areas from which the pathological process spreads through transneuronal, anterograde, or retrograde mechanisms. Answers to these unresolved questions will (1) improve our fragmentary insights into the origin and pathomechanisms of the degenerative process of the polyglutamine disease HD, (2) help to decipher the major principles and mechanisms of its topographical and chronological spread throughout the brain, and (3) provide conclusive explanations for the puzzling phenomenon of the selective vulnerability or resistance of distinct cortical and subcortical brain sites (Rüb et al. 2014b).

3.5 Voxel-Based Morphometric Magnet Resonance Imaging (MRI) In Vivo Studies of the Cerebral Cortex in Presymptomatic Huntington's Disease (HD) Patients

Up to now postmortem studies were mainly performed with patients in the advanced clinical stages of HD. Since postmortem studies do not necessarily reflect the initiation and topographical and chronological progress of HD-related neurodegeneration, *voxel-based morphometric magnetic resonance imaging (MRI)* in vivo *studies* could be considered as the method of choice to fill this gap. The findings of in vivo MRI studies of affected HD patients cover a long-lasting time period of progressive neurodegeneration. Thus, in vivo longitudinal voxel-based morphometric MRI studies in combination with psychological testing of presymptomatic HD gene carriers are presently considered as the gold standard in HD imaging (Tabrizi et al. 2011). One of the final goals of these in vivo studies is the establishment of reliable biomarkers that allow assessing of the progression of the disease and the potential effects of therapeutical interventions on the clinical symptoms and disease course. The initial results of these longitudinal MRI studies of premanifest HD gene carriers and/or early HD patients included progressive global brain atrophy (Tabrizi et al. 2011), progressive cortical volume loss (Nopoulos et al. 2010), worsening atrophy of the striatum (Aylward et al. 2011; Tabrizi et al. 2011), as well as white matter loss in the frontal lobe (Aylward et al. 2011). The progressive changes in the striatum and white matter of the frontal lobes were considered as potential biomarkers for HD (Aylward et al. 2011).

This chapter was written by Helmut Heinsen and Udo Rüb.

Chapter 4
Degeneration of Select Motor and Limbic Nuclei of the Thalamus in Huntington's Disease (HD)

4.1 The Thalamus in Huntington's Disease (HD)

Although early neuropathological studies reported degenerative features in subcortical brain regions of HD patients (e.g., pallidum, subthalamic nucleus, substantia nigra, claustrum, amygdala) including the *thalamus* (e.g., global atrophy and volume loss of the thalamus; thalamic astrogliosis; reduced density of microneurons in its cerebellar territory, the ventrolateral nucleus; astrogliosis in the centromedian and parafascicular nuclei; shrinkage of the centromedian nucleus and its nerve cells) (see Chap. 1) (Borrell-Pagès et al. 2006; Dom et al. 1976; Finkbeiner and Mitra 2008; Lange and Aulich 1986; Lange et al. 1976; McCaughey 1961; McLardy 1948; Pfeiffer 1913; Terplan 1924; Vonsattel 2008; Vonsattel and DiFiglia 1998; Vonsattel et al. 1985; Walker 2007a, b), the involvement of the thalamus and the extent of its degeneration in HD have been neglected in more recent investigations. Therefore, the affection of the thalamic nuclei is currently not among the established degenerative features of HD.

According to current pathoanatomical concepts, the typical distribution pattern of brain neurodegeneration in HD is thought to be the result of an anterograde, retrograde, or transneuronal and topographically highly ordered spread of the underlying pathological process throughout the brain via interconnecting brain fiber tracts. Spreading of this process may involve a *prion-like mechanism* (Brundin et al. 2010; Costanzo and Zurzolo 2013; Goedert et al. 2010; Jucker and Walker 2011; Labbadia and Morimoto 2013; McLardy 1948; Millecamps and Julien 2013; Norrby 2011; Renner and Melki 2014). In view of this possibly *targeted spread of the disease process along anatomical pathways*, it appeared conceivable that brain regions possessing intimate anatomical connections with the affected regions of the cerebral cortex (e.g., prefrontal cortex) and/or the neostriatum may also be among the targets of the neurodegenerative process of HD (Heinsen et al. 1996, 1999; Heinsen and

U. Rüb et al., *The Neuropathology of Huntington's Disease: Classical Findings, Recent Developments and Correlation to Functional Neuroanatomy*,
Advances in Anatomy, Embryology and Cell Biology, Vol. 217,
DOI 10.1007/978-3-319-19285-7_4

Rüb 1997). In view of the *severe degeneration of the neostriatum and the layer-specific neuronal loss in the prefrontal cortex*, it appeared plausible that *the thalamic centromedian-parafascicular complex* (*CM-PF*) *and mediodorsal nucleus* (*MD*) which are interconnected with these cortical and subcortical predilection sites of HD-associated neurodegeneration may also be among the primary or secondary targets of the disease process of HD. Indeed, these theoretical assumptions have been confirmed by recent investigations that applied modern stereological methods (i.e., volume assessments according to the Cavalieri's principle; cell counts with the optical disector) on thick frontal serial tissue sections through the thalamus stained with gallocyanin for neuronal Nissl material and demonstrated the consistent neurodegeneration of both thalamic nuclei in HD patients (Heinsen and Heinsen 1991; Heinsen et al. 1996, 1999).

4.2 Functional Neuroanatomy of the Centromedian-Parafascicular Complex (CM-PF) of the Human Thalamus

The centromedian-parafascicular complex (*CM-PF*) (Fig. 4.1) is a collection of neurons that belong to the caudal intralaminar nuclear group, which is located in the posterior part of the thalamus. The *parafascicular nucleus* (*PF*) (Fig. 4.1) occurs in all mammals and, in contrast to the other thalamic nuclei (e.g., the ventrolateral nucleus), is well conserved during the phylogeny of the primate brain. The *centromedian nucleus* (*CM*) (Fig. 4.1), however, appears later in the evolution, is enlarged in parallel with the disproportionally increase of the putamen, becomes increasingly prominent in primates, and reaches its maximal development in humans (Armstrong 1990; Fenelon et al. 1991; Rapoport 1990; Sadikot and Rymar 2009; Sadikot et al. 1992b). The CM and PF are the source of the main thalamic afferents to the striatum (Fig. 4.2) and together innervate most portions of the striatum in primates in a topographical order. The CM represents a nodal point in the closed reciprocal *neostriatal-pallido-centromedian-neostriatal loop of the basal ganglia*, is the most prominent source of thalamostriatal projections, and is intricately linked to both components of the neostriatum (i.e., putamen and caudate nucleus) by efferent projections. Via these striatofugal projections, the CM is involved in the regulation of the activity of the neostriatal output neurons. The CM receives its main afferents from the internal segment of the pallidum via the lenticular ansa and the lenticular and thalamic fascicles. It is therefore included into the motor basal ganglia-thalamocortical feedback loop, is mainly associated with the *sensorimotor territory of the striatum* (Fig. 2.9), and can serve sensorimotor integrative functions and processing of motor information to the neostriatum (Fenelon et al. 1991; François et al. 1991; Groenewegen 2003; Haber and Calzavara 2009; Heinsen et al. 1996; Nakano et al. 2000; Parent 1990; Parent and Hazrati 1995; Sadikot and Rymar 2009; Sadikot et al. 1992a; Sadikot et al. 1992b; Walker 1982). The PF projects mainly to the *associative territory of the*

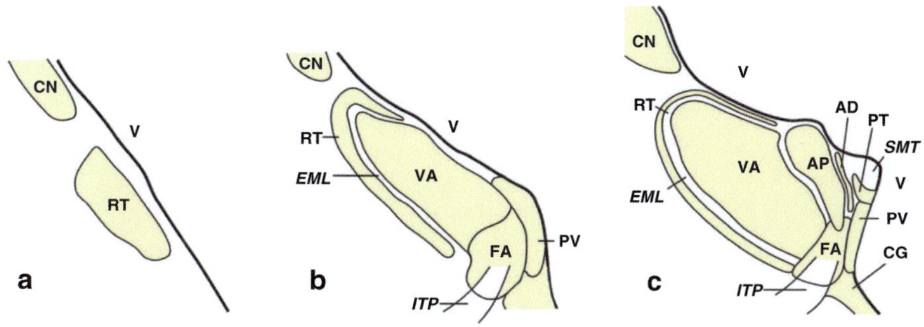

Fig. 4.1 The human thalamus. Schematized frontal sections cut perpendicularly to Forel's intercommissural axis showing the rostrocaudal sequence of the nuclei of the human thalamus. (**a**) Rostral pole of the human thalamus made up by the reticular nucleus (*RT*). (**b**) Frontal section through the rostral pole of the human thalamus with the lateral portion of the reticular (*RT*), ventral anterior (*VA*), fasciculosus (*FA*), and paraventricular nuclei (*PV*). (**c**) The human thalamus with the reticular (*RT*), ventral anterior (*VA*), fasciculosus (*FA*), paraventricular (*PV*), parataenial (*PT*), anterodorsal (*AD*), and anteroprincipal nuclei (*AP*). (**d**) The rostral thalamus at the level of the mammillothalamic tract (*MMT*) showing the reticular (*RT*), ventral lateral (*VL*), ventral anterior (*VA*), central medial (*CEM*), mediodorsal (*MD*), paraventricular (*PV*), parataenial (*PT*), anterodorsal (*AD*), and anteroprincipal nuclei (*AP*). (**e**) The mid-level of the human thalamus comprises the reticular (*RT*), ventral lateral (*VL*), mediodorsal (*MD*), paracentral (*PC*), parafascicular (*PF*), paraventricular (*PV*), cucullar (*CU*), parataenial (*PT*), anterodorsal (*AD*), and anteroprincipal nuclei (*AP*). (**f**) Frontal section through the caudal thalamus at the level of the habenular nuclei (*H*): reticular nucleus (*RT*), lateral geniculate body (*LGB*), inferior nucleus of the pulvinar (*PU i*), lateral posterior nucleus (*LP*), ventral lateral nucleus (*VL*), ventral posterior lateral (*VPL*) and ventral posterior medial nuclei (*VPM*), anterior nucleus of the pulvinar (*PUa*), ventral posterior medial nucleus, parvocellular part (*VPMpc*), centromedian (*CM*), parafascicular (*PF*), subparafascicular (*SPF*), central lateral (*CL*), mediodorsal (*MD*), and laterodorsal nuclei (*LD*). (**g**) The human thalamus at the level of the pretectum with the reticular (*RT*) and lateral posterior nuclei (*LP*), the lateral (*PU l*) and inferior nuclei of the pulvinar (*PU i*), the lateral (*LGB*) and medial geniculate bodies (*MGB*), the medial nucleus of the pulvinar (*PU m*), the central lateral nucleus (*CL*), as well as the limitans-suprageniculate complex (*LI-SG*). (**h**) Caudal pole of the human thalamus with the reticular nucleus (*RT*) and the lateral (*PU l*), inferior (*Pu i*), and medial nuclei (*PU m*) of the pulvinar. The nuclei investigated by means of modern stereological methods in HD (i.e., MD, CM, and PF) are colored in *red* (Modified according to Rüb et al. (2003b), (Figure 1, page 2260); with kind permission from Oxford University Press). Abbreviations: *AD* anterodorsal nucleus, *AP* anteroprincipal nucleus, *ATW* triangular area of Wernicke, *CEM* central medial nucleus, *CG* central gray, *CL* central lateral nucleus, *CM* centromedian nucleus, *CN* caudate nucleus, *CTT* corticotectal tract, *CU* cucullar nucleus, *EML* external medullary lamina, FA fasciculosus nucleus, *H* habenular nuclei, *ITP* inferior thalamic peduncle, *LD* laterodorsal nucleus, *LGB* lateral geniculate body, *LI-SG* limitans-suprageniculate complex, *LP* lateral posterior nucleus, *MD* mediodorsal nucleus, *MGB* medial geniculate body, *MMT* mammillothalamic tract (Vicq d'Azyr), *PC* paracentral nucleus, *PF* parafascicular nucleus, *PR* pretectum, *PT* parataenial nucleus, *PUa* anterior nucleus of the pulvinar, *PU i* inferior nucleus of the pulvinar, *PU l* lateral nucleus of the pulvinar, *PU m* medial nucleus of the pulvinar, *PV* paraventricular nucleus, *RT* reticular nucleus, *SMT* stria medullaris thalami, *SPF* subparafascicular nucleus, *V* ventricle, *VA* ventral anterior nucleus, *VL* ventral lateral nucleus, *VPL* ventral posterior lateral nucleus, *VPM* ventral posterior medial nucleus, *VPMpc* ventral posterior medial nucleus of the thalamus, parvocellular part

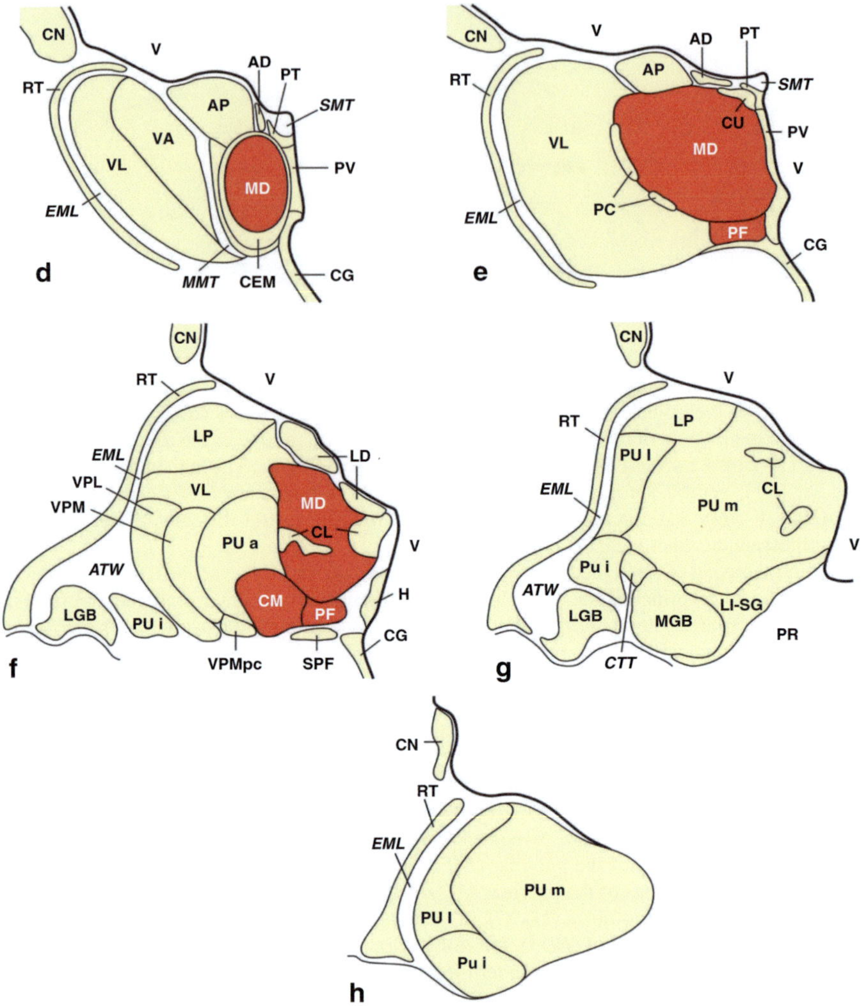

Fig. 4.1 (continued)

striatum (Figs. 2.9 and 4.2) and less abundantly to its *limbic territory* (e.g., accumbens nucleus) (Fig. 2.9). In addition, the PF is reciprocally interconnected with the oculomotor frontal and supplementary eye fields of the cerebral cortex. Owing to its anatomical interconnectivities and its relation with the associative and limbic territories of the striatum and the oculomotor network, the PF is involved in associative-limbic brain functions (i.e., cognitive and emotional processes), as well as oculomotor functions (Fenelon et al. 1991; Heinsen et al. 1996; Huerta et al. 1986; Künzle and Akert 1977; Nakano et al. 2000; Parent 1990; Parent and Hazrati 1995; Sadikot and Rymar 2009; Sadikot et al. 1992a; Sadikot et al. 1992b; Stanton et al. 1988).

The *CM-PF* of the human thalamus is embedded into an extension of the internal medullary lamina and represents the main component of the caudal intralaminar

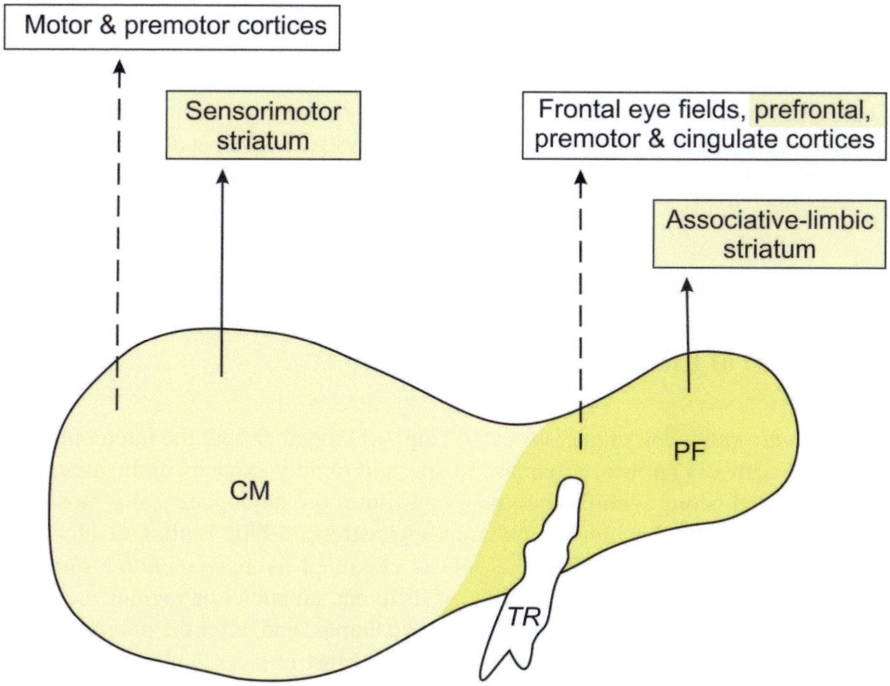

Fig. 4.2 The major projections of the human thalamic centromedian-parafascicular complex (CM-PF) to the cerebral cortex and striatum. Diagram of a frontal section through the human thalamic centromedian (*CM*, light yellow) and parafascicular (*PF*, *yellow*) nuclei with their main efferent outputs to the cerebral cortex (*dashed lines*) and the "sensorimotor," "associative," and "limbic" territories of the striatum (*unbroken lines*). All cortical and subcortical sites depicted in this diagram that have been convincingly shown to sustain neuronal loss in Huntington's disease (HD) are colored either in *yellow* or *light yellow* (Modified according to Sadikot et al. (1992b), (Figure 12, page 156); with kind permission from John Wiley and Sons). Abbreviations: *CM* centromedian nucleus of the thalamus, *PF* parafascicular nucleus of the thalamus, *TR* tractus retroflexus, habenulo-interpeduncular tract

group (Fig. 4.1). The CM-PF is rostrally and dorsally bordered by the large, associative thalamic MD and laterally by the somatosensory ventroposterior medial and the gustatory parvocellular portion of the ventroposterior medial thalamic nuclei. Separated from it by a small medullary lamina, the subparafascicular nucleus lies ventrally to the CM-PF and is integrated into the limbic circuits of the human brain. Medially, the CM-PF abuts on the central gray and merges dorsally with the pulvinar and the limitans-suprageniculate complex, which marks the border between the thalamus and the midbrain pretectum (Figs. 4.1 and 4.4). Although many authors recognized two subnuclei or suggested a tripartite architectonic subdivision of the human CM-PF complex, all investigators of the normal anatomy of this thalamic complex agree that the lateral boundaries of the medially situated PF are complex and interdigitate without sharp transitions with the medial parts of the CM (Fig. 4.4) (Hassler 1982a; Heinsen et al. 1996; Herrero et al. 2002; Hirai and Jones 1989; Jones 1985; Morel et al. 1997; Sadikot and Rymar 2009; Walker 1982).

In Nissl-stained tissue sections through the human thalamus, the two subnuclei of the human CM-PF exhibit marked differences in their nerve cell densities and staining features. The CM displays a lower cell density than the PF and is richer in fibers with strands and cellular islands irregularly arranged within its boundaries (Fig. 4.4) (Hassler 1982a; Heinsen and Heinsen 1991; Heinsen et al. 1996; Herrero et al. 2002; Hirai and Jones 1989; Jones 1985; Morel et al. 1997; Sadikot and Rymar 2009; Walker 1982).

4.3 Functional Neuroanatomy of the Mediodorsal Nucleus (MD) of the Human Thalamus

The *thalamic mediodorsal nucleus* (*MD*) (Fig. 4.1) together with the interconnected prefrontal cortex is known to increase in size and rapidly expands in the ascending phylogeny and, along with the pulvinar, constitutes the phylogenetically most progressed nucleus of the human thalamus (Armstrong 1990; Barbas et al. 1991; Rapoport 1990; Walker 1982). The MD is classified as an *associative thalamic nucleus*, represents a large aggregation of different subnuclei or regions, occupies wide regions in the central part of the human thalamus, and, after the pulvinar, is the largest nuclear complex in the human thalamus. Most intracerebral connections of the MD are established with the *prefrontal cortex* (Figs. 4.1 and 4.3). Owing to its reciprocal interconnections with the prefrontal cortex via efferent (i.e., thalamocortical projections from the MD to the inner granular layer IV of the prefrontal cortex) and afferent pathways (i.e., corticothalamic projections from the inner pyramidal layer V and multiform layer VI of the prefrontal cortex to the MD), the MD is considered as *a major subcortical link to the prefrontal cortex* (Fig. 4.3) (Barbas et al. 1991; Carmichael and Price 1996; Fuster 1989; Giguere and Goldman-Rakic 1988; Haber and Calzavara 2009; Hassler 1982a; Heinsen et al. 1999; Herrero et al. 2002; Jones 1985; Markowitsch 1982; McFarland and Haber 2002; Mesulam 1998; Montoya et al. 2006; Morecraft et al. 1992; Walker 1982; Zilles 1990). Interruptions of these interconnectivities with the prefrontal cortex within the context of *neurosurgical prefrontal leucotomy or lobotomy* performed to relieve the symptoms of neuropsychiatric disorders are well known to be associated with strong degenerative signs in the human MD (Freeman and Watts 1947; Giguere and Goldman-Rakic 1988; Hassler 1982b; Heinsen et al. 1999; Herrero et al. 2002; Meyer et al. 1947). From a functional point of view, the human MD has been implicated in olfactory processing and cognitive, learning, and memory processes, as well as executive brain functions (e.g., goal formation, planning, goal-directed action, self-monitoring, attention, coordination of complex cognition and motor control for effective performance, dynamical focus change between points of fixation) and oculomotor functions (Barbas et al. 1991; Carmichael and Price 1996; Fuster 1989; Giguere and Goldman-Rakic 1988; Harting et al. 1980; Hassler 1982a; Heinsen et al. 1999; Herrero et al. 2002; Huerta et al. 1986; Jones 1985; Künzle and Akert 1977; Markowitsch 1982; McFarland and Haber 2002; Sommer and Wurtz 2004; Stanton et al. 1988).

The human MD is surrounded by a split of the internal medullary lamina and a ring of nuclei located within this lamina (Fig. 4.1). The MD is rostrally bordered by

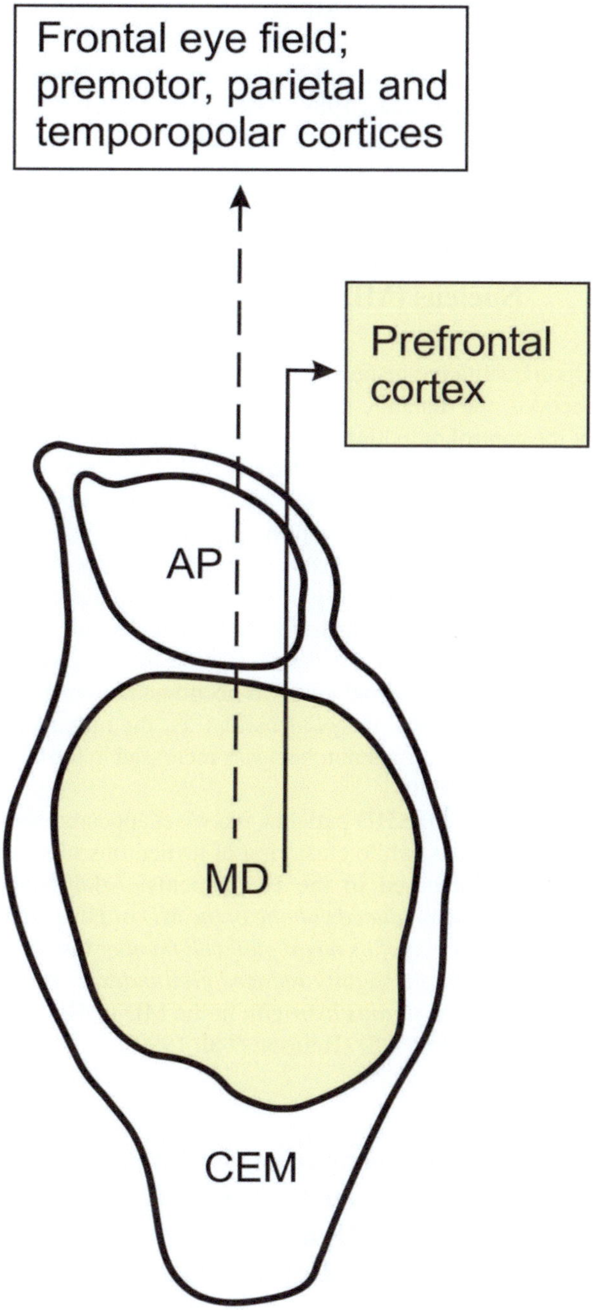

Fig. 4.3 The major projections of the human mediodorsal nucleus (*MD*) to the cerebral cortex. Diagram of a frontal section through the human mediodorsal thalamic nucleus (*MD*) showing its main efferent outputs to the cerebral cortex. The brain structures that are known to undergo neurodegeneration during Huntington's disease (HD) are colored in *light yellow*. Abbreviations: *AP* anteroprincipal nucleus of the thalamus, *CEM* central medial nucleus of the thalamus, *MD* mediodorsal nucleus of the thalamus

the intralaminar central medial nucleus and dorsally by the limbic anteroprincipal and laterodorsal thalamic nuclei, as well as by the small intralaminar cucullar nucleus. The paracentral intralaminar nucleus lies laterally and the midline paraventricular nucleus medially to the MD, while the CM-PF abuts on the MD ventrally

(Fig. 4.1). Caudally, the MD borders on the islands of the intralaminar central lateral nucleus and is replaced by the medial pulvinar and the limitans-suprageniculate complex (Figs. 4.1 and 4.5) (Hassler 1982a; Heinsen et al. 1996; Herrero et al. 2002; Hirai and Jones 1989; Jones 1985; Morel et al. 1997; Walker 1982).

4.4 Degeneration of the Thalamic Centromedian-Parafascicular Complex (CM-PF) and Mediodorsal Nucleus (MD) in Huntington's Disease (HD)

Upon qualitative inspections of gallocyanin-stained thick frontal serial brain tissue sections, the thalamic *CM-PF* of HD patients appeared generally smaller than that of the control individuals. In the HD patients the density of the nerve cells in the PF was more severely reduced than in their CM. *Volume reduction of the CM-PF* of the HD patients was confirmed by quantitative studies using modern stereological methods (i.e., Cavalieri's principle in combination with the optical disector) and revealed an *overall reduction of the absolute neuronal number in the CM-PF* of HD patients by approximately 55 % when compared to age- and gender-matched control individuals (Fig. 4.4). The estimated absolute neuron number in the CM-PF of the HD patients did not correlate with the neuronal number in their neostriatum. Since CM-PF neuronal loss was accompanied *by a considerable GFAP immunoreactive astrogliosis*, the *glial index* (i.e., the numerical ratio between the glial cell number and neuronal number) was increased in the CM-PF of HD patients (Heinsen et al. 1996, 1999).

The *MD* of the HD patients, likewise, appeared smaller than the MD of the control individuals, while clustering of its neurons was less marked, and the density of its fibers increased in the HD patients. Additional stereological investigations demonstrated a *reduced volume of the MD* of HD patients, a *mean nerve cell loss* by 24 %, a *reduction of its mean glial cell number* (astrocytes and oligodendrocytes) by 30 %, as well as a slightly *reduced glial index*. Furthermore, Gallyas silver impregnation revealed *fibrous astroglia* in the MD of HD patients which was absent from the controls (Fig. 4.5) (Heinsen et al. 1999).

4.5 The Involvement of the Thalamic Centromedian-Parafascicular Complex (CM-PF) and Mediodorsal Nucleus (MD) in Huntington's Disease (HD) and its Pathogenetic Implications

Previous investigations of the thalamus suggested various degenerative changes (e.g., global atrophy and volume loss of the thalamus; thalamic astrogliosis; reduced density of microneurons in its cerebellar territory, the ventrolateral nucleus; astrogliosis in the centromedian and parafascicular nuclei; shrinkage of the

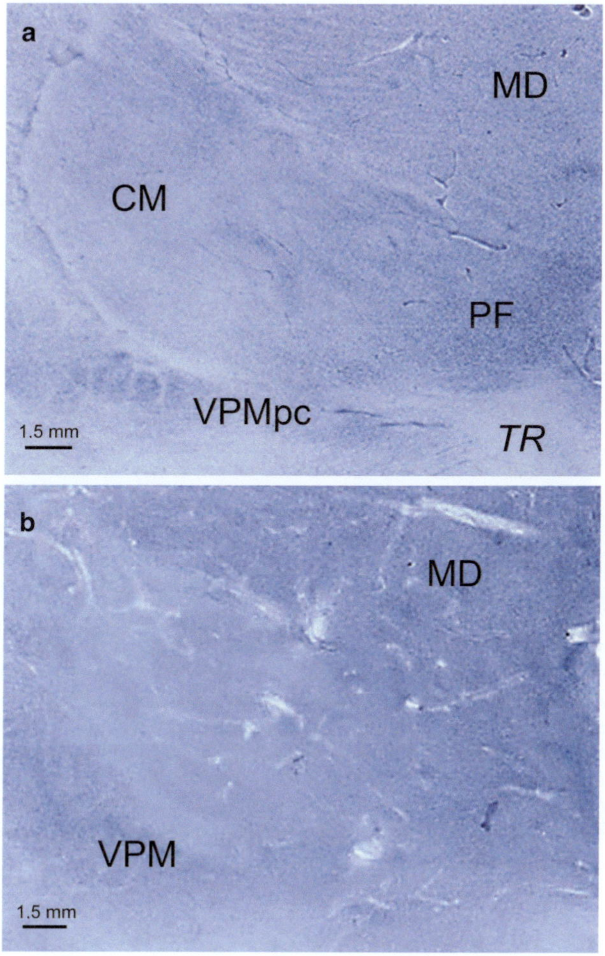

Fig. 4.4 Neuronal loss in the thalamic centromedian-parafascicular complex (CM-PF) in Huntington's disease (HD). (**a**) Frontal section through the midportion of the left centromedian-parafascicular complex (CM-PF) of a representative control individual without neuropsychiatric diseases. (**b**) Reduced volume, pallor, and marked neuronal loss in the CM-PF of a typical Huntington's disease (HD) patient. Pallor and reduction of nerve cells were more severe in the PF than in the CM of the HD patient. The neurodegenerative changes in the thalamus of the HD patient were accompanied by a moderate perivascular edema. (**a**, **b**: gallocyanin staining, 400 μm gelatine sections). (**b**: Modified according to Heinsen et al. (1996), (Figure 3, page 163); with kind permission from Springer Business and Media Center). Abbreviations: *CM* centromedian nucleus of the thalamus, *MD* mediodorsal nucleus of the thalamus, *PF* parafascicular nucleus of the thalamus, *TR* tractus retroflexus (habenulo-interpeduncular tract), *VPM* ventral posterior medial nucleus of the thalamus, *VPMpc* ventral posterior medial nucleus of the thalamus, parvocellular part

centromedian nucleus and its nerve cells) (see Chap. 1) in HD patients (Figs. 1.4) (Borrell-Pagès et al. 2006; Dom et al. 1976; Finkbeiner and Mitra 2008; McCaughey 1961; McLardy 1948; Pfeiffer 1913; Terplan 1924; Vonsattel 2008; Vonsattel and DiFiglia 1998; Vonsattel et al. 1985; Walker 2007a, b). Some of these observations

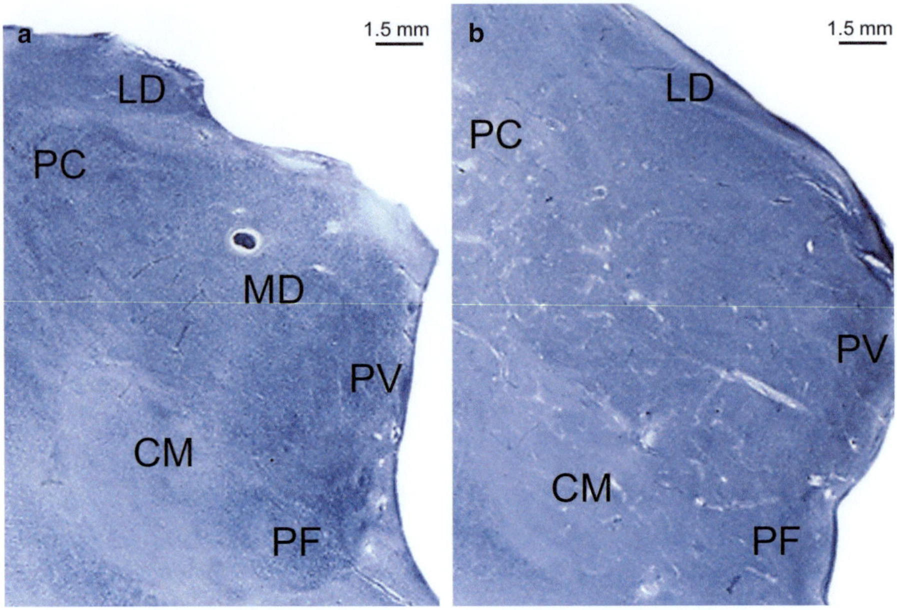

Fig. 4.5 Neuronal loss in the thalamic mediodorsal nucleus (*MD*) in Huntington's disease (HD). (**a**) Frontal section through the caudal third of the left mediodorsal nucleus (*MD*) of a typical control individual. (**b**) Volume reduction, pallor, and marked neuronal loss in the MD of the same Huntington's disease (HD) patient as in Fig. 4.4. The degenerative changes in the thalamic centromedian (*CM*) and parafascicular (PF) nuclei are described in more detail in the legend of Fig. 4.4. (**a**, **b**: gallocyanin staining, 400 µm gelatine sections). Abbreviations: *CM* centromedian nucleus of the thalamus, *LD* laterodorsal nucleus of the thalamus, *MD* mediodorsal nucleus of the thalamus, *PC* paracentral nucleus of the thalamus, *PF* parafascicular nucleus of the thalamus, *PV* paraventricular nucleus of the thalamus

were confirmed and extended by recent morphometrical investigations of select thalamic nuclei of HD patients, which showed that thalamic neurodegeneration in HD is evident not only in its ventrolateral nucleus but also in its CM-PF and MD (Figs. 4.4 and 4.5) (Dom et al. 1976; Heinsen et al. 1996, 1999).

The thalamic CM, PF, and MD are connected via brain fiber tracts with the prominent brain targets of the neurodegenerative process of HD (i.e., neostriatum, prefrontal cortex). The *CM* is included into the motor basal ganglia-thalamocortical feedback loop and via its anatomical interconnectivities that is mainly associated with the *sensorimotor territory of the striatum* (i.e., dorsolateral parts of the precommissural and postcommissural putamen, lateral part of the body of caudate nucleus) (Figs. 2.9, 4.2, and 4.3) (Fenelon et al. 1991; Heinsen et al. 1996; Nakano et al. 2000; Parent 1990; Parent and Hazrati 1995; Sadikot and Rymar 2009; Sadikot et al. 1992a; Sadikot et al. 1992b). The *PF*, in contrast, is more closely related to the *associative* (i.e., the head of the caudate nucleus, medial portions of the body, and the tail of the caudate nucleus, precommissural putamen, ventromedial parts of the postcommissural putamen) *and limbic territories of the striatum* (i.e., accumbens nucleus, ventromedial part of the precommissural putamen) (Figs. 2.9 and 4.2) (Fenelon et al. 1991; Heinsen et al. 1996; Nakano et al. 2000; Parent 1990; Parent and Hazrati 1995; Sadikot and Rymar 2009; Sadikot et al. 1992a; Sadikot et al. 1992b). The MD is reciprocally connected with the *prefrontal cortex* and receives topographically ordered afferents from and emits topographically ordered efferents to this neocortical target of the neurodegenerative process of HD (Fig. 4.3). Thalamocortical projections from the MD terminate in the inner granular layer IV and outer pyramidal layer III of the prefrontal cortex, while nerve cells of the prefrontal multiform layer VI and, to a lesser extent, of the inner pyramidal layer V emit corticothalamic efferents to the thalamic MD (Fig. 4.3) (Barbas et al. 1991; Carmichael and Price 1996; Fuster 1989; Giguere and Goldman-Rakic 1988; Haber and Calzavara 2009; Hassler 1982a; Heinsen et al. 1999; Herrero et al. 2002; Jones 1985; Markowitsch 1982; McFarland and Haber 2002; Mesulam 1998; Morecraft et al. 1992; Walker 1982; Zilles 1990).

To explain the targeting of select cortical and subcortical brain regions and the brain distribution pattern of neuronal demise in HD is still one of the most important and enduring challenges and remains the first-rate goal of current and future HD research. Previously attempted explanations for the high *vulnerability of select cortical and subcortical sites* pointed to the *anatomical interconnectivities* of a given brain region (1) as a crucial factor accounting for their susceptibility for or resistance to the degenerative process of HD and (2) as the possible structural basis for the apparently well-ordered and stereotypical brain progression of this process and the resulting characteristic distribution pattern of neurodegeneration. Although the exact underlying molecular biological mechanisms still remain to be determined, currently many researchers in the field favor the pathomechanistic hypothesis that the degenerative process of the HD-related pathology (1) follows anatomical pathways during its brain evolution, (2) spreads transneuronally via anatomical pathways in predictable and consistent temporal and spatial sequences, and (3) therefore results in a specific distribution pattern of neuronal loss that mirrors the anatomical interconnectivities of involved brain sites (Labbadia and Morimoto 2013; Renner and Melki 2014; Rüb et al. 2014b). Degeneration of the CM-PF and MD in HD is in agreement with these theoretical considerations and supports the current concept that the *anatomical interconnectivities* of a given brain region can account for its susceptibility for or resistance to the degenerative process of HD (Labbadia and Morimoto 2013). Considering (1) that all three functionally defined portions of the striatum may sooner or later be subjected to the degenerative process of HD and (2) that intimate interconnectivities exist between the *CM* and the *sensorimotor territory of the striatum*, as well as between the *PF* and the *associative and limbic territories of the striatum*, damage to these caudal intralaminar thalamic nuclei during HD could represent the long-distance effect of a retrograde spread of the degenerative process of HD that primarily occurs in the striatum and gradually affects its different components. Alternatively damage to these caudal intralaminar thalamic nuclei could constitute an initial degenerative event that secondarily triggers degeneration in the associated territories of the striatum via an anterograde expansion (Heinsen et al. 1996; Imarisio et al. 2008). Furthermore, in view (1) of the repeatedly documented neuronal loss in the outer and inner pyramidal layers III and V, as well as in the multiform layer VI of the *prefrontal cortex* of HD patients, (2) of the thalamocortical projections from the *MD* to the inner granular layer IV and outer pyramidal layer III of the prefrontal cortex, and (3) of the corticothalamic afferents to the MD originating from prefrontal cortex multiform layer VI and, to a lesser extent, from its inner pyramidal layer V, the apparent intactness of the inner granular layer IV strongly points to well-preserved thalamocortical projections to the prefrontal cortex originating from the MD. Since damage to the outer pyramidal layer III of the prefrontal cortex, likewise, may not be linked to impaired thalamocortical projections from the MD, this cortical damage (1) may reflect the anatomical interconnectivities of the prefrontal cortex with the striatum via efferent projections, (2) may constitute a primary cortical degenerative event in the HD-related disease process, or (3) may be secondarily induced by diseased nerve cells in the initially affected striatum and retrogradely conveyed to the prefrontal cortex via fiber tracts (Heinsen et al. 1999; Imarisio et al. 2008). Furthermore, considering the corticothalamic projections of the nerve cells of the inner pyramidal layer V and multiform layer VI of the prefrontal cortex to the thalamic MD, the degeneration of these neuronal layers during HD could be interpreted (1) as a primary degenerative event in the cerebral cortex which secondarily triggers MD degeneration via corticothalamic projections or (2) as the pathological outcome of the retrogradely transmitted disease process emanating from the earlier affected thalamic MD (Heinsen et al. 1999).

Astrocytes are known to secrete neurotrophic factors and cytokines and are activated in response to almost all types of brain injury. In their activated state, astrocytes upregulate the glial fibrillary acidic protein (GFAP), participate in the phagocytosis of cell debris and in the formation of glial scars, and are therefore regarded as sensitive indicators of any kind of brain damage (Eng and Ghirnikar 1994; Eng et al. 2000). In view of the scavenger function of reactive astrocytes in diverse pathological conditions, the detection of *immunoreactive or silver-impregnated astrocytes in the CM-PF or the MD* of HD patients favors the view that the surviving astroglial cells become reactive during HD to phagocytose and form a glial scar that compensates tissue loss after nerve cell degeneration (Heinsen et al. 1996, 1999).

4.6 Functional Implications of the Involvement of the Thalamic Centromedian-Parafascicular Complex (CM-PF) and Mediodorsal Nucleus (MD) in Huntington's Disease (HD)

The estimated neuronal loss of approximately 55 % in the *CM-PF* of HD patients together with that in interconnected subcortical brain sites most likely is associated with functional impairments. The CM receives cortical afferents from the motor, premotor, supplementary, and cingulate motor cortical areas, as well as from the somatosensory cortex, is intimately interconnected with the *sensorimotor territory of the striatum*, represents a nodal point of the *basal ganglia-thalamocortical circuit*, and is therefore crucial for the regulation of the activity of the neostriatal output neurons and can serve integration of sensorimotor functions and processing of motor information to the neostriatum (Figs. 2.9 and 4.2) (Fenelon et al. 1991; Heinsen et al. 1996; Nakano et al. 2000; Parent 1990; Parent and Hazrati 1995; Sadikot and Rymar 2009; Sadikot et al. 1992a; Sadikot et al. 1992b.

In contrast to the CM, the *PF* is mainly linked with the *associative territory of the striatum* (Figs. 2.9 and 4.2) and, although less strongly, with its *limbic territory*. In addition, the PF is also reciprocally interconnected with the *oculomotor frontal and supplementary eye fields of the cerebral cortex* (Fig. 4.2) (Fenelon et al. 1991; Heinsen et al. 1996; Huerta et al. 1986; Künzle and Akert 1977; Nakano et al. 2000; Parent 1990; Parent and Hazrati 1995; Sadikot and Rymar 2009; Sadikot et al. 1992a; Sadikot et al. 1992b; Stanton et al. 1988). Since the PF via these anatomical relationships is involved in the performance of associative-limbic and oculomotor functions, its damage most likely contributes to the *cognitive, psychiatric, emotional, and oculomotor symptoms* that evolve in HD patients (Heinsen et al. 1996).

The MD *constitutes a major subcortical link to the prefrontal cortex*, is integrated into the *limbic and oculomotor circuits of the human brain*, and among others subserves cognitive, learning, and memory processes and executive and oculomotor functions (Fig. 4.3) (Barbas et al. 1991; Carmichael and Price 1996; Fuster 1989; Giguere and Goldman-Rakic 1988; Harting et al. 1980; Hassler 1982a; Heinsen et al. 1999; Herrero et al. 2002; Huerta et al. 1986; Jones 1985; Künzle and Akert 1977; Markowitsch 1982; McFarland and Haber 2002; Montoya et al. 2006; Sommer and Wurtz 2004; Stanton et al. 1988; Walker 1982; Zilles 1990). Accordingly, neuronal loss in the MD may also contribute to *cognitive, mnestic, psychiatric, and emotional disease symptoms of HD patients*. Since the MD is also integrated into the oculomotor circuits of the human brain, degeneration of the MD, along with damage to the thalamic PF and the premotor nuclei of the oculomotor brainstem network (see Sects. 4.2, 4.4, 4.5, 4.6, 6.6, 7.2, and 7.3), could also be involved in the pathogenesis of *oculomotor disturbances* that occur in HD patients and in their at-risk relatives (Heinsen et al. 1996, 1999).

This chapter was written by Udo Rüb.

Chapter 5
Consistent and Widespread Degeneration of the Cerebellum in Huntington's Disease (HD)

5.1 Clinical Symptoms Pointing to an Involvement of the Cerebellum in Huntington's Disease (HD)

The autosomal dominantly inherited polyglutaminopathy causing Huntington's disease (HD) is a currently untreatable and rare neuropsychiatric disorder with an estimated prevalence in Europe and the USA of 4–8:100,000 (Andrew et al. 1993; Finkbeiner and Mitra 2008; Harper 1992; Labbadia and Morimoto 2013; Margolis and Ross 2003; Ortega et al. 2007; Rüb et al. 2009; Schapira et al. 2014; Schulte and Littleton 2011; Tanner and Goldman 1994; Vonsattel and DiFiglia 1998; Walker 2007a, b). The symptoms of adult-onset HD commonly begin around the age of 40 years with progressive cognitive impairments and motor symptoms often designated as clumsiness, tremor, balance trouble, or jerkiness. Although choreatic movements may also be among the early and progressive HD symptoms, they plateau and disappear in the advanced clinical stages. During the late clinical stage, additional somatomotor (i.e., bradykinesia, akinesia, dystonia, hypotonia, rigidity, dysarthria, dysphagia) and oculomotor symptoms, visual and executive dysfunctions, personality changes, psychiatric disturbances (e.g., depression, schizophrenia-like symptoms), electrophysiological abnormalities, and an unintended, severe, and unexplained weight loss may occur (Andrew et al. 1993; Aziz et al. 2008; Borrell-Pagès et al. 2006; Bruyn et al. 1979: Deuschl et al. 1989; Ellenberger et al. 1978; Gil and Rego 2008; Heinsen et al. 1994; Hennerici et al. 1985; Imarisio et al. 2008; Josiassen et al. 1984; Knikou 2008; Kremer et al. 1992; Lasker and Zee 1997; Leigh and Zee 2006; Li and Conforti 2013; Margolis and Ross 2003; McLeod 1969; Misiaszek 2003; Myers 2004; Petersen et al. 2005; Ross and Tabrizi 2011; Rüb et al. 2009, 2013a, 2014a, b; The Huntington's disease Collaborative Research Group 1993; Vonsattel 2008; Vonsattel and DiFiglia 1998; Vonsattel et al. 1985; Walker 2007a, b).

U. Rüb et al., *The Neuropathology of Huntington's Disease: Classical Findings, Recent Developments and Correlation to Functional Neuroanatomy*,
Advances in Anatomy, Embryology and Cell Biology, Vol. 217,
DOI 10.1007/978-3-319-19285-7_5

The gradual worsening *degeneration of the striatum* has been for a long time used to explain the spectrum of HD symptoms, including oculomotor and choreatic symptoms (see Chaps. 1, 2, and 3) (Andrew et al. 1993; Borrell-Pagès et al. 2006; Braak and Braak 1992a, b; De la Monte et al. 1988; Estrada-Sanchez and Rebec 2013; Heinsen et al. 1994, 1996, 1999, Heinsen and Rüb 1997; Imarisio et al. 2008; Lange and Aulich 1986; Lange et al. 1976; Margolis and Ross 2003; Mink 1996; Myers et al. 1988; Ortega et al. 2007; Rüb et al. 2009; 2013a, 2014a, b; Vonsattel 2008; Vonsattel and DiFiglia 1998; Vonsattel et al. 1985; Walker 2007a, b). A variety of symptoms (i.e., dysarthria, impaired rapid alternating movements and fine motor skills, postural and gait instability, ataxia, falls), however, pointed to *cerebellar damage* in HD patients. Since the cerebellum was only fragmentarily investigated in adult HD patients, its affection during HD has been controversially discussed for several decades (Bruyn et al. 1979; Busse et al. 2009; Fennema-Notestine et al. 2004; Grimbergen et al. 2008; Jeste et al. 1984; Knikou 2008; Koller and Trimble 1985; Kremer et al. 1992; Margolis and Ross 2003; Rodda 1981; Rosas et al. 2002, 2003; Rüb et al. 2013a; Vonsattel 2008; Vonsattel and DiFiglia 1998; Vonsattel et al. 1985; Walker 2007a, b).

5.2 Neurodegeneration of the Cerebellum in Huntington's Disease (HD)

Recently, the first systematic and comprehensive investigation of the cerebellum in clinically diagnosed and genetically confirmed HD patients clearly revealed its consistent involvement in HD patients with Vonsattel grades 2–4 of striatal atrophy (Figs. 2.11, 5.1, 5.2, 5.3, 5.4, and 5.5) (Rüb et al. 2013a). *Macroscopically*, the brains of the symptomatic HD patients showing extra-cerebellar symptoms (i.e., choreatic movements, bradykinesia, dysphagia, weight loss, executive dysfunctions, personality changes, cognitive decline) and/or cerebellar symptoms (i.e., unsteady and broad-based gait, gait imbalance, impaired gait coordination, dysarthria, slowed, smooth pursuit eye movements) displayed consistent *atrophic changes of the cerebellum* (i.e., reduction of its entire volume and of the surface area of its arbor vitae; atrophy of the lobules of the anterior and posterior lobes; widened primary fissure) (Fig. 2.11) (Rüb et al. 2013a).

Upon light microscopical examination, a *widespread neuronal loss* was detected *in* the *cerebellar cortex and all deep cerebellar nuclei* of all HD patients studied (Figs. 5.2, 5.3, 5.4, and 5.5). Neuronal loss in the *cerebellar cortex* was assessed in *pigment-Nissl*-stained as well as in *calbindin*-immunostained cerebellar tissue sections and neuronal loss in the deep cerebellar nuclei in pigment-Nissl-stained tissue sections through the atrophic HD cerebella (Figs. 5.2, 5.3, 5.4, and 5.5) (Rüb et al. 2013a). Along with other proteins, calbindin is among the known calcium-binding proteins that participate in the cell signaling pathways of the second messenger calcium and represents an acknowledged, selective, and highly reliable immunocytochemical marker for the cerebellar Purkinje cells. In the cerebellum, rapid firing

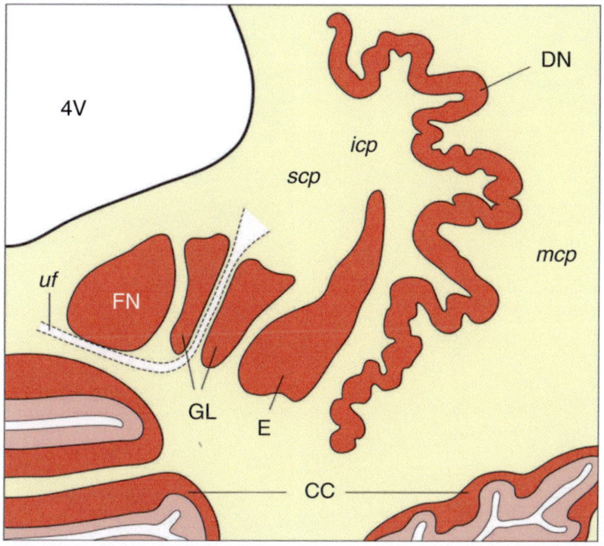

Fig. 5.1 The human cerebellum. Schematized horizontal section through the right human cerebellum showing the cerebellar cortex (*CC*) with its three neuronal layers (i.e., molecular, Purkinje, and granular cell layers). Embedded in the deep cerebellar white matter reside the cerebellar peduncles (*icp* inferior cerebellar peduncle, *mcp* medial cerebellar peduncle, *scp superior* cerebellar peduncle) and the four cerebellar nuclei (the curvated band of the *DN* dentate nucleus; *E* the cigar-shaped emboliform nucleus; *GL* the globose nucleus split by the uncinate fascicle (*uf*) into multiple portions; *FN* the roof ridge-like fastigial nucleus) (Modified according to Scherzed et al. (2012), (Figure 2, page 752); with kind permission from Springer Science and Business Media). Abbreviations: *CC* cerebellar cortex, *DN* dentate nucleus, *E* emboliform nucleus, *FN* fastigial nucleus, *GL* globose nucleus, *icp* inferior cerebellar peduncle, *mcp* medial cerebellar peduncle, *scp* superior cerebellar peduncle, *uf* uncinate fascicle, *4V* fourth ventricle

neurons experience high levels of calcium influx. Thus, tight calcium regulation is essential to avoid rapid neuronal death due to excessive excitatory stimulation (Bastianelli 2003). The large GABAergic Purkinje cells serve as the sole output of the cerebellar cortex, project to the deep cerebellar and brainstem nuclei, and are at 100 % immunoreactive for the calcium-binding protein calbindin (Bastianelli 2003). To highlight cerebellar Purkinje cells, calbindin immunostaining is more reliable than the traditional Nissl staining procedures, which were developed by the German neuroscientist Franz Nissl (1860–1919), since in the human cerebellum these neurons are only weakly stained by the Nissl technique (Whitney et al. 2008).

Investigation of the cerebellum of HD patients disclosed a consistent and selective loss of Purkinje cells, while the molecular and granular cell layers typically were spared (Fig. 5.2) (Rüb et al. 2013a). *Loss of Purkinje cells* was found in all lobules of the vermis and cerebellar hemispheres. It was not evenly distributed over the affected cerebellar lobules, but occurred at circumscribed predilection sites, which were intermingled between well-preserved and unremarkable segments of the cerebellar lobules. Surviving Purkinje cells often appeared *ballooned or*

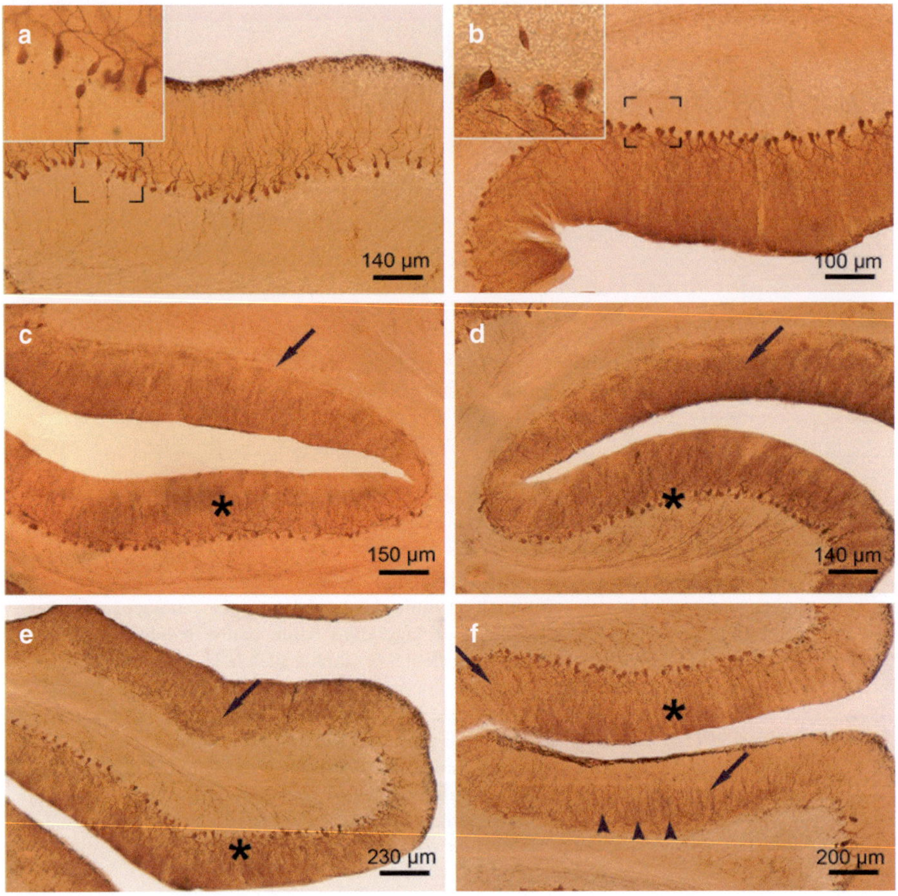

Fig. 5.2 The cerebellar Purkinje cell layer in Huntington's disease (HD). (**a**) Sagittal section through the vermal portion of the cerebellar lobule X of a Huntington's disease (HD) patient with Vonsattel grade 2 of neostriatal atrophy. The Purkinje cell layer is intact and shows no loss of these cerebellar output neurons. Framed area is depicted at higher magnification at *top left* and shows a torpedo-like axonal inclusion. (**b**) Sagittal section through the vermal portion of the cerebellar lobule V of the same HD patient. Although the Purkinje cell layer is well preserved, the Purkinje cells showed torpedo-like inclusions in their axons (framed area and *inset* at *top left*). (**c**) Hemispheral portion of the cerebellar lobule VI of an HD patient with Vonsattel grade 4 of neostriatal atrophy: intact segments of the Purkinje cell layer (*asterisk*) are intermingled with sections showing complete Purkinje cell loss (*arrow*). (**d**) Hemispheral portion of lobule VI of a HD patient with Vonsattel grade 2 of neostriatal atrophy: Intact segments of the Purkinje cell layer (*asterisk*) and degenerated segments with serious loss of Purkinje cells (*arrow*). (**e**) Hemispheral portion of lobule VI of a HD patient with Vonsattel grade 2 of neostriatal atrophy: along with intact Purkinje cell layer segments (*asterisk*), portions with severe loss of Purkinje cells (*arrow*) prevail. (**f**) Hemispheral portion of cerebellar lobule IX of an HD patient with Vonsattel grade 2 of neostriatal atrophy: portions with no Purkinje cell loss, but dendritic alterations (*asterisk*) alternate with regions affected by severe loss of Purkinje nerve cells (*arrow*). Note the remaining and pathologically altered dendritic trees (*arrowheads*) (**a**–**f** Anti-calbindin immunostaining, 100 μm polyethylene glycol sections) (Reprinted from Rüb et al. (2013a), (Figure 2, page 168); with kind permission from John Wiley and Sons)

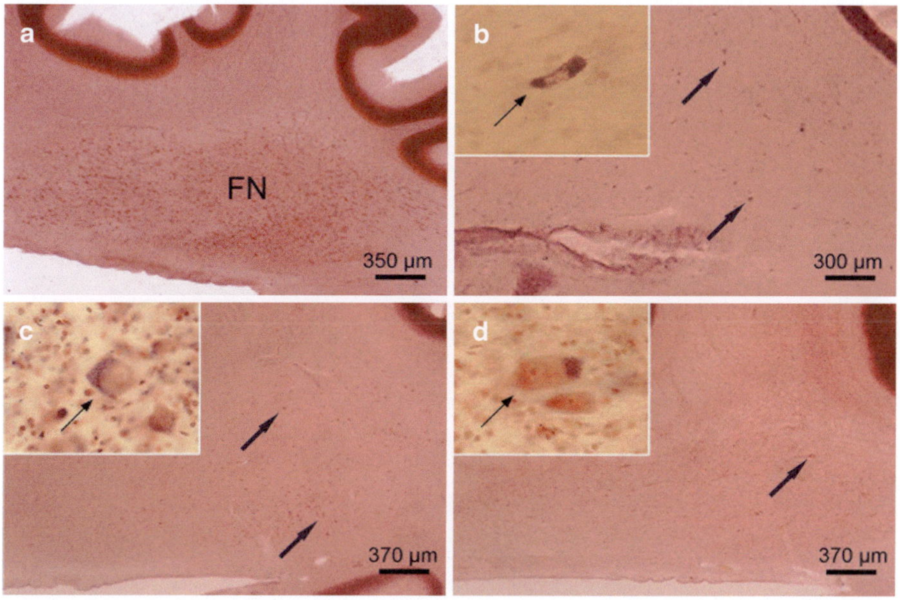

Fig. 5.3 The cerebellar fastigial nucleus (*FN*) in Huntington's disease (HD). (**a**) Sagittal section through the right cerebellum of a representative control individual depicting the roof ridge-like fastigial nucleus (*FN*). (**b**) Dramatic neuronal loss in the FN of a representative Huntington's disease (HD) patient with Vonsattel grade 2 of neostriatal atrophy. *Arrows* point to remaining FN nerve cells, and *inset* shows severely shrunken surviving FN nerve cell (*thin arrow*). (**c**) Severe neuronal loss in the FN of an HD patient with Vonsattel grade 2 of neostriatal atrophy. Remaining FN nerve cells are marked by *arrows* and *inset* shows ballooned FN nerve cell (*thin arrow*). These swollen nerve cells are characterized by an enlarged cytoplasm with a basophilic center, peripherally displaced nucleus, Nissl substance, and lipofuscin granules. (**d**) Seriously affected FN of an HD patient with Vonsattel grade 4 of neostriatal atrophy. *Arrow* indicates a surviving FN nerve cell, and *inset* shows ballooned surviving FN nerve cells (*thin arrow*) (**a–d** aldehyde-fuchsin Darrow red staining, 100 µm polyethylene glycol sections) (Reprinted from Rüb et al. (2013a), (Figure 4, page 170); with kind permission from John Wiley and Sons). Abbreviations: *FN* fastigial nucleus

shrunken, and their axons in some instances harbored *torpedo-like inclusions* (Fig. 5.2) (Rüb et al. 2013a).

In addition to the cerebellar Purkinje cell layer, the four *deep cerebellar nuclei* (i.e., *fastigial*, *globose*, *emboliform*, *and dentate nuclei*) (Fig. 5.1) embedded into the cerebellar white matter also underwent a marked to severe neuronal loss during HD (Figs. 5.3, 5.4, and 5.5) (Rüb et al. 2013a). Among these deep cerebellar nuclei, the roof ridge-like *fastigial nucleus* commonly was most severely affected and nearly devoid of nerve cells. Most residual neurons appeared ballooned or shrunken. These *swollen or ballooned fastigial nerve cells* had a rounded and massively enlarged cytoplasm with a homogeneous central basophilic substance. Their Nissl substance, lipofuscin granules, and flattened nucleus were concentrated along the cytoplasmic membrane and thus mimicked central chromatolysis.

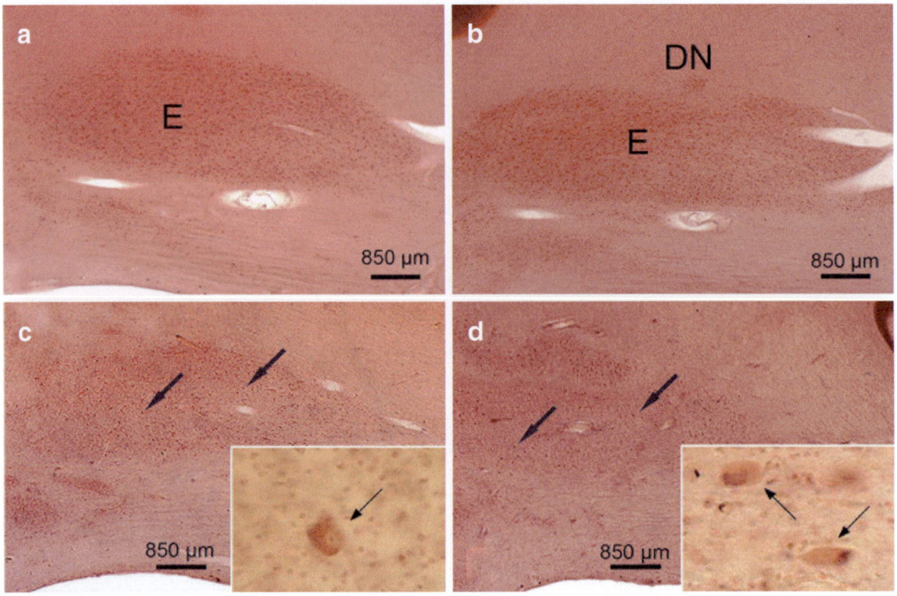

Fig. 5.4 The cerebellar emboliform nucleus (*E*) in Huntington's disease (HD). (**a**) Sagittal section through the right cerebellum of a representative control individual depicting the medial portion of the cigar-shaped emboliform nucleus (*E*). (**b**) Sagittal section through the right cerebellum of a typical control individual depicting the lateral portion of the cigar-shaped E and the most medial portion of the curvated band of the dentate nucleus (*DN*). (**c**) Marked neuronal loss in the medial portion of the E of a Huntington's disease (HD) patient with Vonsattel grade 3 of neostriatal atrophy. *Arrows* point to remaining E nerve cells, and *inset* shows remaining ballooned E nerve cell (*thin arrow*) with bloated cytoplasm, space-consuming basophilic center, peripherally displaced nucleus, Nissl material, and lipofuscin granules. (**d**) Severe neuronal loss in the lateral portion of the E and marked neuronal drop-out in the most medial portion of the DN of an HD patient with Vonsattel grade 2 of neostriatal atrophy. *Arrows* point to remaining E nerve cells, and *inset* shows ballooned surviving E nerve cells (*thin arrows*) (**a**–**d**: aldehyde-fuchsin Darrow red staining, 100 µm polyethylene glycol sections) (Reprinted from Rüb et al. (2013a), (Figure 5, page 173); with kind permission from John Wiley and Sons). Abbreviations: *DN* dentate nucleus, *E* emboliform nucleus

By contrast, the *shrunken fastigial neurons* displayed a pale, slender, and arrowhead-like cytoplasm. Their central and pale nucleus was broader and stretched the abnormal cytoplasm, which was nearly devoid of Nissl substance (Fig. 5.3) (Rüb et al. 2013a).

The cigar-shaped *emboliform nucleus*, likewise, was consistently affected in HD patients and underwent a severe neuronal demise. As with the affected fastigial nucleus, the majority of surviving emboliform neurons were ballooned or shrunken (Fig. 5.4) (Rüb et al. 2013a). Although neuronal loss was consistently present in the *globose and dentate nuclei* of most of the HD patients studied, it was less severe than in their fastigial or emboliform nuclei. However, also the globose and dentate nuclei showed a clearly reduced nerve cell density and numerous surrounding lipofuscin granules in the neuropil (Fig. 5.5) (Rüb et al. 2013a). *Lipofuscin granules*

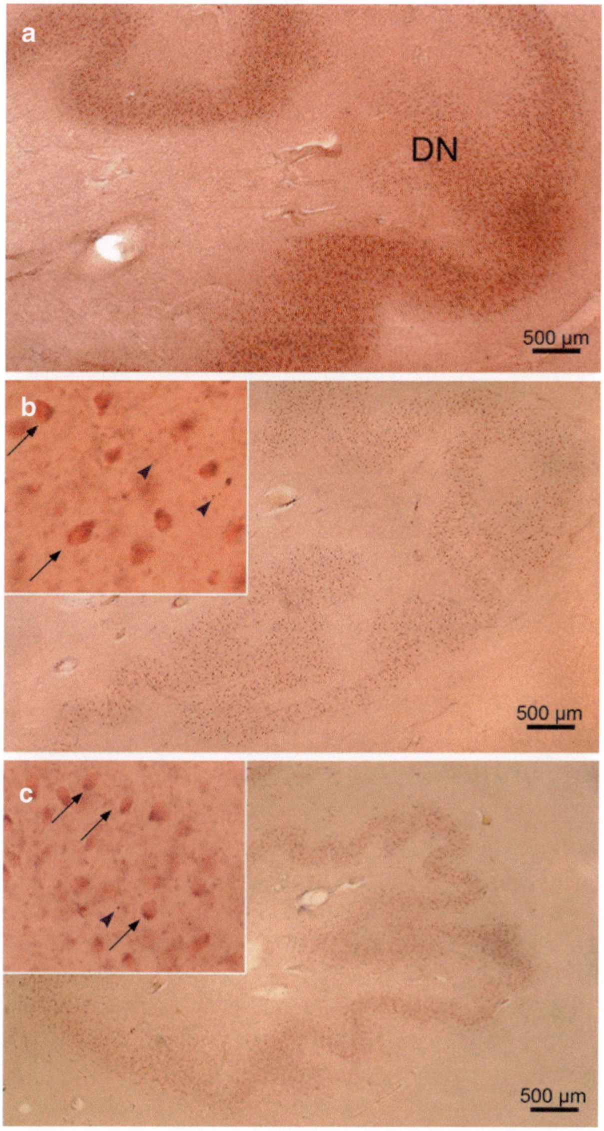

Fig. 5.5 The cerebellar dentate nucleus (*DN*) in Huntington's disease (HD). (**a**) Sagittal section through the right cerebellum of a control individual with no history of neuropsychiatric diseases depicting the folded dentate nucleus (*DN*). (**b**) Marked neuronal loss of the DN of a Huntington's disease (HD) patient with Vonsattel grade 2 of neostriatal atrophy. *Inset* at *top left* points to surviving DN nerve cells: ballooned nerve cells with bloated cytoplasm and space-consuming basophilic center, peripherally displaced nucleus, Nissl material, and lipofuscin granules (*thin arrows*), and extraneuronal lipofuscin granules (*arrowheads*). (**c**) Considerable neuronal loss in the DN of an HD patient with Vonsattel grade 3 of neostriatal atrophy. *Inset* at *top left* shows remaining nerve cells: ballooned nerve cells (*thin arrows*), extraneuronal lipofuscin granules (*arrowhead*) (**a–c**: aldehyde-fuchsin Darrow red staining, 100 polyethylene glycol sections) (Reprinted from Rüb et al. (2013a), (Figure 4, page 174); with kind permission from John Wiley and Sons). Abbreviations: *DN* dentate nucleus

are normal cytoplasmic constituents of nerve and glial cells and are structurally stable postmortem even after delayed or suboptimal fixation. Owing to the low affinity of glial and the high affinity of neuronal lipofuscin granules to the aldehyde-fuchsin dye, lipofuscin granules in the neuropil can serve as reliable markers of the former position of lipofuscin-laden neuronal perikarya that have disappeared from brain tissue (Braak et al. 2003b; Rüb et al. 2001, 2003a, 2004b, 2008a, b, 2009, 2013b; Scherzed et al. 2012).

Neither Purkinje cell loss nor neuronal loss in the deep cerebellar nuclei of the HD patients was correlated with the length of the CAG repeats in the mutated HD alleles, age at HD onset, duration of HD, age at death, or the Vonsattel grade of striatal atrophy (Rüb et al. 2013a).

Despite careful and systematic investigations of serial cerebellar tissue sections, no remaining nerve cells in the cerebellar cortex and deep cerebellar nuclei were found to display morphological features (i.e., chromatin condensation, nuclear fragmentation, apoptotic bodies) associated with the occurrence of *classical apoptosis* (Graeber and Moran 2002; Rüb et al. 2013a).

Reactive astrocytes immunopositive for the glial fibrillary acidic protein (GFAP) could be observed in all cerebellar components of HD patients. They were most frequent in the cerebellar white matter, but also consistently present in the three layers of the cerebellar cortex, as well as in all four deep cerebellar nuclei (Rüb et al. 2013a).

5.3 The Functional Relevance of Cerebellar Affection in Huntington's Disease (HD)

Since the cerebellum was only fragmentarily investigated in adult HD patients, its involvement during HD has been controversially discussed for several decades (see Chap. 1) (Bruyn et al. 1979; Fennema-Notestine et al. 2004; Jeste et al. 1984; Koller and Trimble 1985; Kremer et al. 1992; Margolis and Ross 2003; Rodda 1981; Rosas et al. 2003; Rüb et al. 2013a; Vonsattel 2008; Vonsattel and DiFiglia 1998; Vonsattel et al. 1985; Walker 2007a, b). This situation has changed since the first comprehensive study of the cerebellum was published (Rüb et al. 2013a). This recent study clearly demonstrated consistent pathological changes involving the cerebellum of HD patients with Vonsattel grade 2–4 striatal atrophy. It also showed unequivocally that *cerebellar neurodegeneration* in HD is not confined to the Purkinje cell layer and/or the dentate nucleus as assumed previously by some authors, but includes macroscopically visible cerebellar atrophy, widespread loss of Purkinje cells, as well as a considerable and consistent demise of nerve cells in all four deep cerebellar nuclei (i.e., fastigial, globose, emboliform, and dentate nucleus) (Figs. 2.9, 5.1, 5.2, 5.3, 5.4, and 5.5) (Bruyn et al. 1979; Jeste et al. 1984; Rodda 1981; Rosas et al. 2003; Rüb et al. 2013a; Vonsattel and DiFiglia 1998; Vonsattel et al. 1985). These new and consistent neuropathological findings conform to the results from recent in vivo *magnetic resonance imaging* (*MRI*) *studies* (Fennema-Notestine et al. 2004; Rosas et al. 2003; Rüb et al. 2013a) and are highly reminiscent to the cerebellar findings in *autosomal dominantly inherited polyglutamine ataxias* (e.g., spinocerebellar ataxias types 1, 2, and 3). All findings indicate that the brain neuropathology of the polyglutamine disease HD is more closely related to these polyglutamine ataxic disorders than previously thought and underscore the contemporary concept of *HD as a multisystem degenerative disease* affecting the striatum, cerebral neo- and allocortex, thalamus, brainstem, and cerebellum (see Chaps. 1, 2, and 3) (Atkin and Paulson 2014; Borrell-Pagès et al. 2006; Braak and Braak 1992a, b; Bruyn et al. 1979; de la Monte et al. 1988; Dom et al. 1976; Dunlap 1927; Estrada-Sanchez and Rebec 2013; Fennema-Notestine et al. 2004; Ferrante et al. 1987; Finkbeiner and Mitra 2008; Hedreen et al. 1991; Heinsen et al. 1992, 1994, 1996, 1999; Heinsen and Rüb 1997; Hoche et al. 2008; Imarisio et al. 2008; Jeste et al. 1984; Koller and Trimble 1985; Kremer et al. 1992; Lange 1981; Lange et al. 76; Lastres-Becker et al. 2008; Margolis and Ross 2003; McCaughey 1961; Myers et al. 1988; Rodda 1981; Rosas et al. 2003; Rüb et al. 2008a, 2013a, b; Scherzed et al. 2012; Schulte and Littleton 2011; Seidel et al. 2012; Sieradzan and Mann 2001; Sotrel et al. 1991; Valera et al. 2005; Vogt and Vogt 1920, 1942; Vonsattel 2008; Vonsattel and DiFiglia 1998; Vonsattel et al. 1985; Walker 2007a, b).

Since atrophy of the cerebellum and neuronal loss in the cerebellar Purkinje cell layer and in the deep cerebellar nuclei was also present in HD patients with initial and still mild Vonsattel grade 2 of striatal atrophy, damage to the cerebellum may be among the early events during the pathological process of HD and does not inevitably develop simultaneously with, but independently from, the well-known degeneration of the neostriatum. Considering the possibility of such an independent development of cerebellar damage, it is conceivable that the neostriatum is not the sole induction site of the underlying pathological process of HD as believed for a long time, but that this process is characterized by a *polytopic or multifocal onset* (Heinsen et al. 1994; Lange and Aulich 1986, Rüb et al. 2013a; Vonsattel 2008).

In humans the cerebellum plays a crucial role for the initiation and termination of movements, for the smoothly integrated coordination and fine adjustment of limb movements, trunk and eye movements, as well as for the regulation of posture and balance, which is achieved by adjusting the programs of the major descending cortical and subcortical motor brain systems (i.e., pyramidal system, corticopontine tract) (Fig. 5.6 and 5.7) (Ghez 1991; Gilman 1992; Rüb et al. 2013a; Scherzed et al. 2012; Voogd 2003; Voogd et al. 1990). The cerebellum is organized into three functionally defined longitudinal zones with different anatomical connections with the brain and spinal cord: vestibulocerebellum, spinocerebellum, and pontocerebellum. The *vestibulocerebellum or archicerebellum* is represented by the flocculonodular lobe, the *spinocerebellum or paleocerebellum* comprises the vermis and paravermal zone, while the *pontocerebellum or neocerebellum* involves the lateral part of the cerebellar hemispheres. The vestibulocerebellum is concerned with oculomotor functions, governs body equilibrium during gait and stance by controlling axial muscles via its afferent and efferent connections, and is also involved in the maintenance of body equilibrium during stance and gait. The spinocerebellum plays a major role for the ongoing execution of limb movements, control of movement velocity, force, and pattern of muscle activity, while the pontocerebellum is crucial

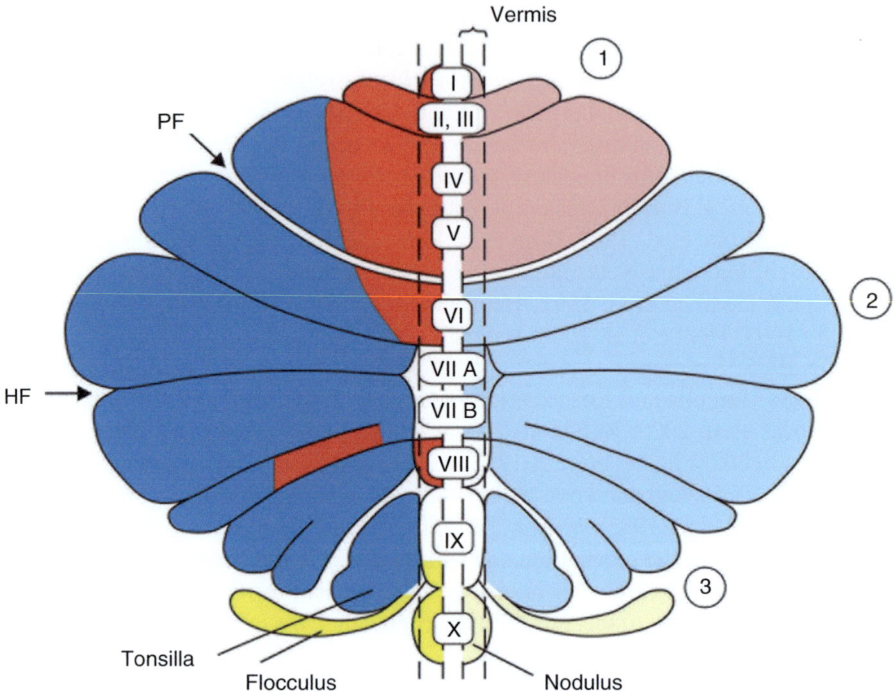

Fig. 5.6 The three phylogenetically and functionally defined longitudinal zones of the human cerebellum. The human cerebellum is composed of three longitudinal functional zones with different anatomical interconnectivities with the brain and the spinal cord: vestibulocerebellum (*yellow*) or archicerebellum (*light yellow*), spinocerebellum (*red*) or paleocerebellum (*light red*), and pontocerebellum (*blue*) or neocerebellum (*light blue*) (Modified according to Benninghoff and Drenckhahn (2004), (Figure 12.6.-16, page 394); with kind permission from Elsevier GmbH, Urban & Fischer). Abbreviations: *Lobules I, II, and III* lingula and central lobule, *Lobules IV and V* culmen, *Lobule VI* declive, *lobule VII* tuber and folium, *Lobule VIII* pyramis, *Lobule IX* uvula, *Lobule X* nodule, *HF* horizontal fissure, *PF* primary fissure, *1* anterior lobe, *2* posterior lobe, *3* flocculonodular lobe

for the planning, initiating, and terminating of movements and for postural control, oculomotor functions, muscle tone, as well as speech production (Figs. 5.6 and 5.7) (Ghez 1991; Gilman 1992; Rüb et al. 2013a; Scherzed et al. 2012; Voogd 2003; Voogd et al. 1990).

Damage to the three *phylogenetically and functionally defined longitudinal cerebellar zones* has been linked to the occurrence of distinct characteristic clinical syndromes: (1) damage to the *flocculonodular lobe* is associated with disturbances of equilibrium, gait ataxia, wide-based standing, and gait position; (2) damage to the *vermal and paravermal portion of the cerebellum and the fastigial nucleus* is correlated with dysarthria, impairments of axial and truncal control, gait and stance ataxia, hypotonia, imbalance, as well as with a reduced activity of gamma motoneurons of the spinal cord; while (3) damage to the *hemispheral portion of the cerebellum and*

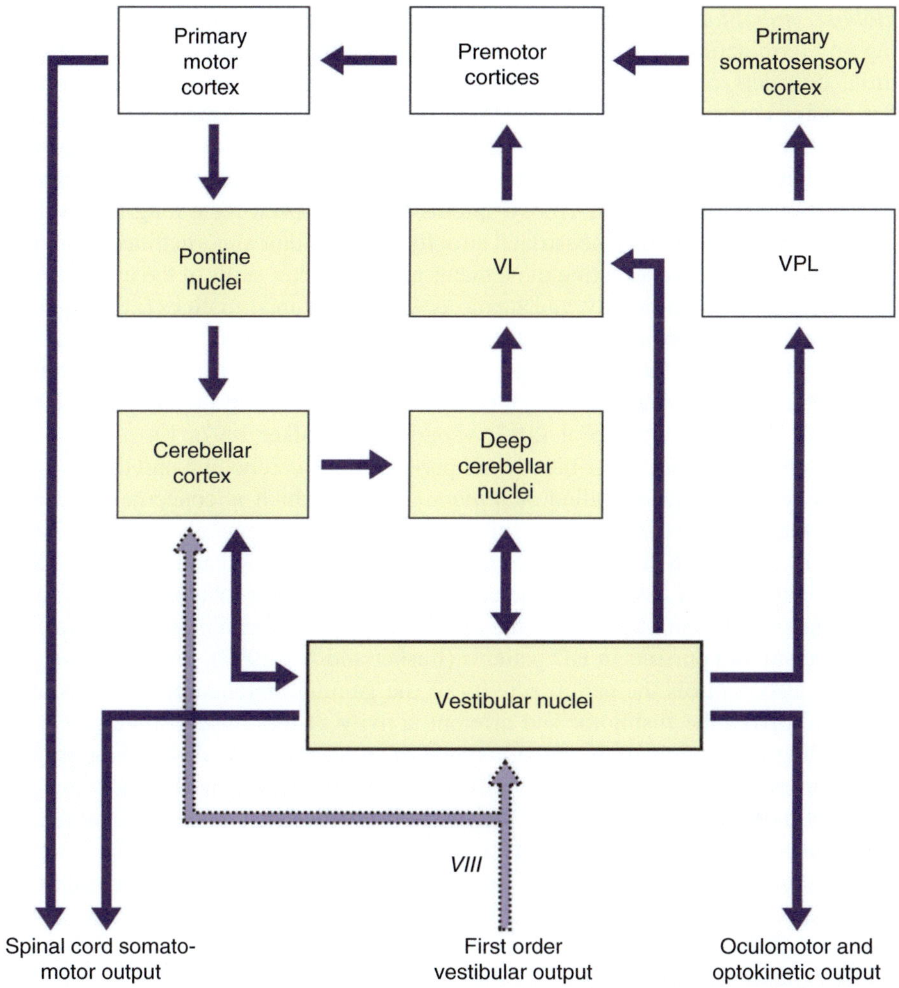

Fig. 5.7 The anatomical interconnectivities of the cerebellum and vestibular and thalamic nuclei in the human brain. First-order vestibular output from the vestibular sense organs reaches the second-order neurons of the superior (SU), medial (MV), lateral (LV), and spinal vestibular nuclei (SV) via the vestibular ganglion and the vestibulocochlear nerve (*VIII*). These projections to the vestibular nuclei give off collaterals (i.e., mossy fibers) to the cortex of the vestibulocerebellum or archicerebellum. The second-order vestibular fibers transmit the output from the vestibular nuclei destined for the vestibulocerebellum, oculomotor brainstem nuclei, and spinal cord. Vestibular information from the MV and SU are also conveyed to the ventral posterior lateral (*VPL*) and ventral lateral thalamic nuclei (*VL*) and, from there, to the primary somatosensory and premotor cortices. Brain regions of these cerebellar and precerebellar neural circuits that have been convincingly shown to undergo neurodegeneration during Huntington's disease (HD) are colored in *yellow* (Reprinted from Rüb et al. (2004a), (Figure 7, page 411); with kind permission from John Wiley and Sons). Abbreviations: *VL* ventrolateral nucleus of the thalamus, *VPL* ventral posterior lateral nucleus of the thalamus, *VIII* vestibulocochlear nerve

globose, *emboliform and/or dentate nuclei* comes along with ataxia, delay in initiating and terminating movements, hypotonia, dysmetria, dysdiadochokinesia, dysarthria, disorders of movement coordination, and aberrant programs of the major descending cortical and subcortical motor brain systems (i.e., pyramidal system, corticopontine tract). Accordingly, damage to the cerebellum together with the lesions in the interconnected ventrolateral thalamic nucleus (Dom et al. 1976) offers suitable explanations for a variety of HD symptoms that have been for a long time solely assigned to the well-known neostriatal atrophy: minor motor abnormalities including impairments of rapid alternating movements and fine motor skills of the upper limbs, dysarthria, postural instability, and ataxia, as well as gait abnormalities (i.e., impaired gait balance, lateral swaying, broad-based gait and stance, falls) (Busse et al. 2009; Ghez 1991; Gilman 1992; Grimbergen et al. 2008; Koller and Trimble 1985; Kremer et al. 1992; Margolis and Ross 2003; Ortega et al. 2007; Robinson and Fuchs 2001; Rüb et al. 2013a; Scherzed et al. 2012; Voogd 2003; Walker 2007a, b).

The lobules VI and VII of the vermal portion of the cerebellum and the fastigial nucleus belong to the so-called *oculomotor vermis* which is concerned with the control and adaption of smooth pursuit eye movements (Horn et al. 1999; Robinson and Fuchs 2001; Rüb et al. 2013a; Scherzed et al. 2012; Voogd and Barmack 2006; Voogd et al. 2010). Loss of Purkinje cells in these cerebellar lobules and the severe neuronal loss in the fastigial nucleus, therefore, provide plausible explanations for impaired smooth pursuits in HD patients (Lasker and Zee 1997). Since the *fastigial nucleus* also controls the neural activity of the gamma motoneurons in the spinal cord and drives the fusimotor and afferent activity of the muscle spindles via its descending projections in the reticulospinal tract, loss of these descending projections can explain why HD patients suffer from muscle hypotonia and show impairments of the *Hoffmann reflex* (*H-reflex*). The observation that this reflex is impaired early during the clinical course of HD again supports the concept that the cerebellum is affected at an early stage of the pathological process of HD (Deuschl et al. 1989; Knikou 2008; McLeod 1969; Misiaszek 2003; Oepen and Hillesheimer 1981; Rüb et al. 2013a, Yanagisawa 1992).

This chapter was written by Udo Rüb and Horst-Werner Korf.

Chapter 6
Elucidation of the Role of the Premotor Oculomotor Brainstem Nuclei in the Pathogenesis of Oculomotor Dysfunctions in Huntington's Disease (HD)

6.1 Types of Eye Movements that Guarantee Stabilization of the Images of Objects of the Visual World Onto the Central Foveal Region of the Retina

The stabilization of the images of objects of the visual world onto the central foveal region of the retina enables a clear and stable view of our environment which is achieved by *seven different types of eye movements* (Table 6.1) (Leigh and Kennard 2004; Leigh and Zee 2006; Rüb et al. 2008b, 2009). The principal aim of these different types of eye movements is either to stabilize gaze to hold images steadily on the central foveal region of the retina or to shift gaze and bring images of objects of the visual world to the retina's fovea independent of head movements. The eye movements are generated by distinct and widely separated oculomotor circuits, which involve a variety of premotor oculomotor brainstem nuclei and converge at the level of the cranial nerve nuclei concerned with eye movements (i.e., oculomotor, trochlear, and abducens nuclei). The oculomotor functions are traditionally subdivided on the basis of how they aid vision: saccades, smooth pursuits, vergence, vestibulo-ocular reflex, optokinetic nystagmus, fixation, and gaze holding (Table 6.1) (Büttner and Büttner-Ennever 2006; Büttner-Ennever and Horn 1997; Leigh and Kennard 2004; Leigh and Zee 2006; Rüb et al. 2008b, 2009).

Saccades represent conjugate eye movements that bring new images of objects of interest onto the central foveal region of the retina and can be voluntary or present as fast phases of the *vestibulo-ocular reflex* and the *optokinetic nystagmus* (Table 6.1). In contrast, *smooth pursuits* are comparatively slow conjugate eye movements that track and hold the image of a small moving visual target on the fovea (Table 6.1). *Vergence* is accomplished by disconjunctive eye movements toward (i.e., convergence) or away from each other (i.e., divergence). These eye movements direct the fovea of both eyes to a single visual object simultaneously

U. Rüb et al., *The Neuropathology of Huntington's Disease: Classical Findings, Recent Developments and Correlation to Functional Neuroanatomy*,
Advances in Anatomy, Embryology and Cell Biology, Vol. 217,
DOI 10.1007/978-3-319-19285-7_6

Table 6.1 Types of eye movements generated by the human oculomotor system and their dysfunctions in Huntington's disease (HD)

Type of eye movement	Normal function-dysfunctions in Huntington's disease (HD)
Saccades	Fast conjugate eye movements that bring new images of objects of interest onto the central foveal region of the retina; saccades can be voluntary or present as fast phases of vestibulo-ocular reflex and optokinetic nystagmus
Saccades in HD	Initiation deficits and slowing and restricted range of vertical and horizontal saccades; loss of vertical and horizontal saccades
Smooth pursuits	Conjugate eye movements that track and hold the image of a small moving visual target on the fovea
Smooth pursuits in HD	Velocity reduction and interruption by inappropriate saccades or square wave jerks
Vergence	Disconjunctive eye movements toward (i.e., convergence) or away from each other (i.e., divergence) which direct the fovea of both eyes at a single visual object simultaneously and guarantee nearing of visual objects and stereoscopic vision
Vergence in HD	Inability to converge
Vestibulo-ocular reflex	Conjugate eye movements with a slow tracking phase and a rapid resetting phase which hold images of the seen world steady on the fovea during brief head rotations or translations
Vestibulo-ocular reflex in HD	Reduced gain of the slow component
Optokinetic nystagmus	Combined conjugate slow tracking pursuit movements and quick repositioning saccades which track and keep the images of large moving visual targets on the fovea
Optokinetic nystagmus in HD	Reduced gain of the slow or rapid component; loss of optokinetic nystagmus
Fixation	Keeps the image of stationary objects of the visual world on the fovea by minimizing ocular drifts and suppresses saccades that turn the fovea away from the object of interest
Fixation in HD	Steady fixation impaired by unwanted saccades
Gaze holding	Stabilizes images of objects of the visual world subsequent to gaze shifting and permits stable eye position between eye movements
Gaze holding in HD	Vertical and horizontal gaze holding consistently intact (even in the late clinical HD stages)

Modified according to Rüb et al. (2009), (Table 1, page 7); with kind permission from John Wiley and Sons

and thus guarantee nearing of visual objects and stereoscopic vision (Table 6.1). The *vestibulo-ocular reflex* involves conjugate eye movements with a slow tracking phase and a rapid resetting phase which hold images of the seen world steady on the fovea during brief head rotations or translations (Table 6.1). The *optokinetic nystagmus* comprises a combination of conjugate slow tracking pursuit movements and quick repositioning saccades that track and keep the images of large moving visual targets on the fovea (Table 6.1). During *fixation* the image of stationary objects of the visual world is kept on the fovea by minimizing ocular drifts and suppression of unwanted saccades that turn the fovea away from the object of interest (Table 6.1).

Gaze holding guarantees stabilization of images of objects of the visual world subsequent to gaze shifting and permits stable eye position between eye movements (Table 6.1) (Büttner and Büttner-Ennever 2006; Büttner-Ennever and Horn 1997; Leigh and Kennard 2004; Leigh and Zee 2006; Rüb et al. 2008b, 2009).

6.2 Oculomotor Dysfunctions in Huntington's Disease (HD)

Oculomotor dysfunctions are among the common and most impressive clinical features of HD and are widely used to establish the clinical diagnosis. In the advanced clinical stages affected HD patients may show a broad range of eye movement abnormalities, including impairments of saccades, smooth pursuits, vergence, vestibulo-ocular reaction, optokinetic nystagmus, and fixation (Table 6.1) (Beenen et al. 1986; Blekher et al. 2004, 2006; Bollen et al. 1986; Kennard and Lueck 1989; Lasker and Zee 1997; Leigh and Zee 2006; Leigh et al. 1983; Margolis and Ross 2003; Oepen and Ostertag 1981; Oepen et al. 1981, 1985; Rüb et al. 2009; Walker 2007a, b).

Impairments of saccades represent the most prominent and frequent symptoms in HD patients and manifest as problems with initiating, with slowing, or with a restricted range of horizontal and/or vertical saccades. Initiation deficits of vertical and horizontal saccades are very frequent disease signs of HD patients and may already be present in preclinical HD gene carriers and in the early clinical phase of HD. Owing to these initiation deficits, affected patients often can start saccades only with an associated head thrust or blink. In addition, the vertical and horizontal saccades of HD patients can also be slowed, hypo-, or hypermetric, restricted in range, or completely lost. In the majority of HD patients, slowed and range-limited saccades occur more frequently in the vertical than in the horizontal plane. As with deficits of saccadic initiation, slowed saccades may already be present in presymptomatic HD gene carriers and may emerge early during the course of HD (Table 6.1) (Avanzini et al. 1979; Beenen et al. 1986; Blekher et al. 2004, 2006; Bollen et al. 1986; Collewijn et al. 1988; Golding et al. 2006; Kirkwood et al. 2000; Lasker and Zee 1997; Lasker et al. 1987, 1988; Leigh et al. 1983; Oepen and Ostertag 1981; Oepen et al. 1981, 1985; Peltsch et al. 2008; Rüb et al. 2009; Winograd-Gurvich et al. 2003). *Deficits of vertical and horizontal smooth pursuits* emerge comparatively late during the course of HD and pertain predominantly to velocity reduction and an interruption by inappropriate saccades or square wave jerks (Table 6.1) (Avanzini et al. 1979; Beenen et al. 1986; Collewijn et al. 1988; Kennard and Lueck 1989; Lasker and Zee 1997; Leigh and Zee 2006; Leigh et al. 1983; Rüb et al. 2009; Starr 1967; Winograd-Gurvich et al. 2003). Some HD patients may also show an *inability to converge*, a reduced gain of the slow component of the *vestibulo-ocular reaction*, as well as impairments of the *optokinetic nystagmus* (i.e., reduced gain of the slow or rapid component, loss of the optokinetic nystagmus) (Table 6.1) (Beenen et al. 1986; Blekher et al. 2004, 2006; Fielding et al. 2004; Golding et al. 2006; Kennard and Lueck 1989; Kirkwood et al. 2000; Lasker and Zee 1997; Leigh and

Zee 2006; Leigh et al. 1983; Leopold et al. 1982; Oepen et al. 1981, 1985; Rüb et al. 2009; Starr 1967). *Impaired steady fixation* is also a frequent, early, and characteristic oculomotor dysfunction of HD patients, whereby affected patients are unable to suppress unwanted saccades during fixation and therefore cannot maintain steady fixation (Table 6.1) (Blekher et al. 2004, 2006; Collewijn et al. 1988; Kennard and Lueck 1989; Lasker and Zee 1997; Leigh and Zee 2006; Leigh et al. 1983; Peltsch et al. 2008; Rüb et al. 2009). In contrast to the six other types of eye movements, *vertical and horizontal gaze holding* is consistently intact in HD even in its late clinical stages (Table 6.1) (Rüb et al. 2009).

The occurrence of the oculomotor symptoms also pointed to an involvement of the *premotor oculomotor brainstem network* in HD and underscored the necessity of detailed brainstem studies (Bruyn et al. 1979; Koeppen 1989; Kremer et al. 1992; Lasker and Zee 1997; Leigh et al. 1983, 1985; Rüb et al. 2009; Vonsattel 2008; Vonsattel and DiFiglia 1998; Vonsattel et al. 1985). However no systematic analysis of this complex functional system could yet be performed in the brains of HD patients. Thus, the exact contribution of the nuclei of this network to the pathogenesis of oculomotor symptoms in HD remained unclear for many decades rendering our knowledge regarding the spectrum of morphological brain alterations underlying the various oculomotor impairments of HD patient fragmentary (Rüb et al. 2009).

6.3 The Human Premotor Oculomotor Brainstem Network

The *human premotor oculomotor brainstem network* (1) involves neural circuits subserving the generation of saccades, smooth pursuits, vergence, the vestibulo-ocular reflex, optokinetic nystagmus, fixation, as well as gaze holding, (2) generates and modifies different physiological features of slow and rapid eye movements (i.e., initiation, velocity, accuracy), and (3) steers the three *oculomotor cranial nerve nuclei* (i.e., oculomotor, trochlear, and abducens nuclei) by means of direct or indirect projections (Büttner and Büttner-Ennever 2006; Büttner-Ennever and Horn 1997; Rüb et al. 2008a, 2009).

Except that subserving vergence, much has been learned in the last two decades regarding the functional neuroanatomy of the neural circuits crucial for the generation of saccades, smooth pursuits, the vestibulo-ocular reflex, and optokinetic nystagmus, as well as gaze holding. The oculomotor circuits are widely separated structurally, generate different types of eye movements relatively independently, involve various premotor oculomotor brainstem nuclei, and converge only at the level of the motoneurons of the oculomotor cranial nerve nuclei (Büttner and Büttner-Ennever 2006; Büttner-Ennever and Horn 1997; Rüb et al. 2008b,2009). The combined application of cytoarchitectonical criteria, histochemical and immunocytochemical techniques, as well as advanced tracing methods enabled the identification of a number of important premotor oculomotor nuclei in the brainstem of nonhuman primates and humans. Prominent examples for these nuclei are the *rostral interstitial nucleus of the medial longitudinal fascicle* (*riMLF*), the *area of excitatory burst neurons for horizontal saccades* (*EBN*), *the raphe interpositus nucleus* (*RIP*) of the paramedian pontine reticular formation, as well as the *dorsal paragigantocellular reticular nucleus* (*DPGI*) (Figs. 6.1, 6.2, and 6.3) (Büttner and Büttner-Ennever 2006; Büttner-Ennever and Horn 2004; Büttner-Ennever et al. 1982, 1988; Horn 2006; Horn and Büttner-Ennever 1998; Horn et al. 1994, 1995; Rüb et al. 2008b, 2009).

The human premotor oculomotor brainstem nuclei are located in the midbrain and pontine tegmentum. They are intimately associated with the *medial longitudinal fascicle* (*MLF*) and represent comparatively small gray components with distinctive cytoarchitectonical features (Figs. 6.1, 6.2, and 6.3). Like other human motor or premotor brainstem nuclei, their nerve cells typically contain only small amounts of cytoplasmic lipofuscin granules. These characteristic cyto- and pigmentoarchitectonical features enable the straightforward and reliable identification of the human premotor oculomotor brainstem nuclei in unconventional thick serial tissue sections stained for neuronal Nissl material and lipofuscin pigment (*pigment-Nissl staining according to Braak*) (Figs. 6.4 and 6.5). This technical approach is therefore well suited to complement and to replace more labor-intensive and expensive histochemical or immunocytochemical staining methods in the field of the anatomy and pathoanatomy of the human oculomotor system (Braak et al. 2003b; Rüb et al. 2001, 2003a, 2004a, b, 2008a, b, 2009, 2013b; Scherzed et al. 2012). Owing to their optical superposition, *unconventional thick serial brain tissue sections* display considerably more nerve cells than the conventional thin sections routinely used by neuropathologists and therefore facilitate the identification of discrete, even minute neuroanatomical components such as the premotor oculomotor brainstem nuclei. The combined staining for neuronal Nissl material (with Darrow red) and lipofuscin pigment (with aldehyde-fuchsin) takes advantage of pigmentoarchitectonical information to facilitate detailed anatomical analyses as well as the evaluation of the exact topographic localization of neurodegeneration. Aldehyde-fuchsin stains neuronal cytoplasmic *lipofuscin granules* prominently, which are structurally stable and can be identified even after delayed fixation. Since their extraneuronal presence indicates the previous position of lost neurons, the presence of these granules in the brain neuropil can serve as a marker for the demise of lipofuscin-containing nerve cells (Braak et al. 2003b; Rüb et al. 2001, 2003a, 2004a, b, 2008a, b, 2009, 2013b; Scherzed et al. 2012).

Some of the premotor oculomotor brainstem nuclei involved in the generation of saccades have only been recently identified and described in studies on the brains of nonhuman primates and humans (i.e., riMLF, EBN, and RIP) (Figs. 6.1, 6.2, 6.3, and 6.4). These studies represent milestones in our understanding of the *functional neuroanatomy of the human premotor oculomotor brainstem network* and provided the base for detailed clinicopathological correlations in human neurodegenerative diseases. The highly interconnected premotor oculomotor brainstem nuclei are made up of the rostral interstitial nucleus of the medial longitudinal fascicle (riMLF), the interstitial nucleus of Cajal (C), the superior colliculus (SC), the reticulotegmental nucleus of the pons (RTTG; nucleus Bechterew), the superior (SUV)

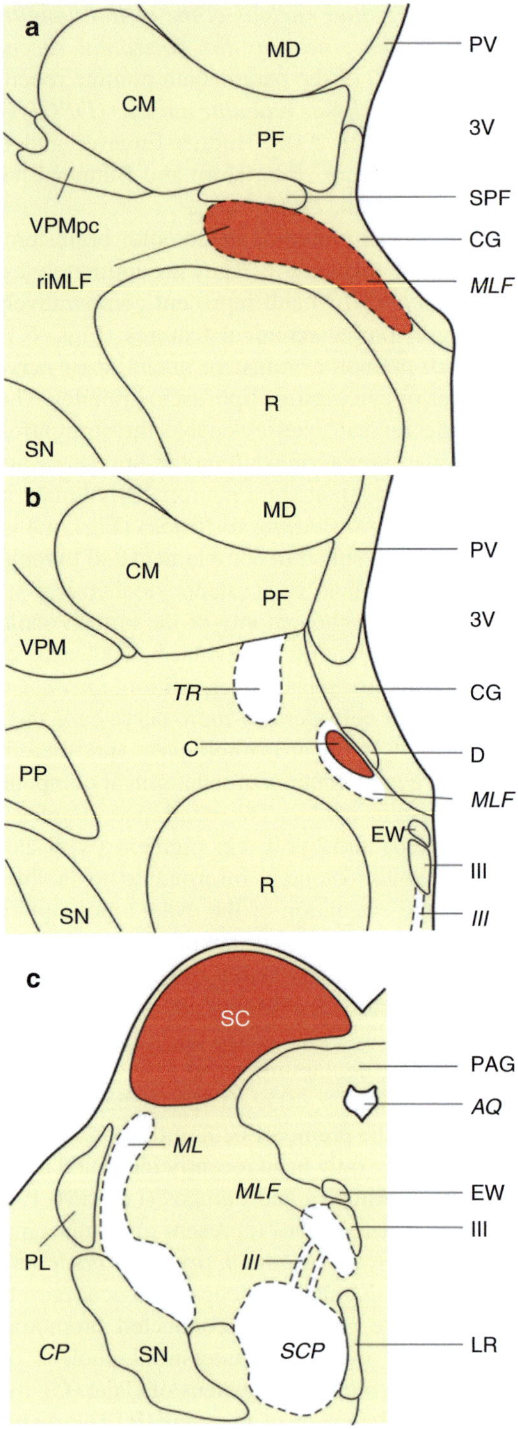

Fig. 6.1 The premotor oculomotor nuclei of the human midbrain. (**a**) Frontal section cut perpendicularly to the intercommissural axis of Forel showing the rostral midbrain with the medial longitudinal fascicle (*MLF*) and the rostral interstitial nucleus of the medial longitudinal fascicle (*riMLF*). (**b**) Frontal section cut perpendicularly to the intercommissural axis of Forel showing the rostral midbrain with the *MLF* and the interstitial nucleus of Cajal (*C*). (**c**) Schematized horizontal section through the caudal midbrain with the superior colliculus (*SC*) (Modified according to Rüb et al. (2008b), (Figure 1, page 170); with kind permission from Springer Science and Business Media). Abbreviations: *AQ* aqueduct, *C* interstitial nucleus of Cajal, *CG* central gray, *CM* centromedian nucleus of the thalamus, *CP* cerebral peduncle, *D* nucleus of Darkschewitsch, *EW* Edinger-Westphal nucleus, *LR* linear raphe nucleus, *MD* mediodorsal nucleus of the thalamus, *ML* medial lemniscus, *MLF* medial longitudinal fascicle, *PAG* periaqueductal gray, *PF* parafascicular nucleus of the thalamus, *PL* paralemniscal nucleus, *PP* peripeduncular nucleus, *PV* paraventricular nucleus of the thalamus, *R* red nucleus, *riMLF* rostral interstitial nucleus of the medial longitudinal fascicle, *SC* superior colliculus, *SCP* superior cerebellar peduncle, *SN* substantia nigra, *SPF* subparafascicular nucleus of the thalamus, *TR* tractus retroflexus (habenulo-interpeduncular tract), *VPM* ventral posterior medial nucleus of the thalamus, *VPMpc* ventral posterior medial nucleus of the thalamus, parvocellular part, III oculomotor nucleus, *III* oculomotor nerve, *3V* third ventricle

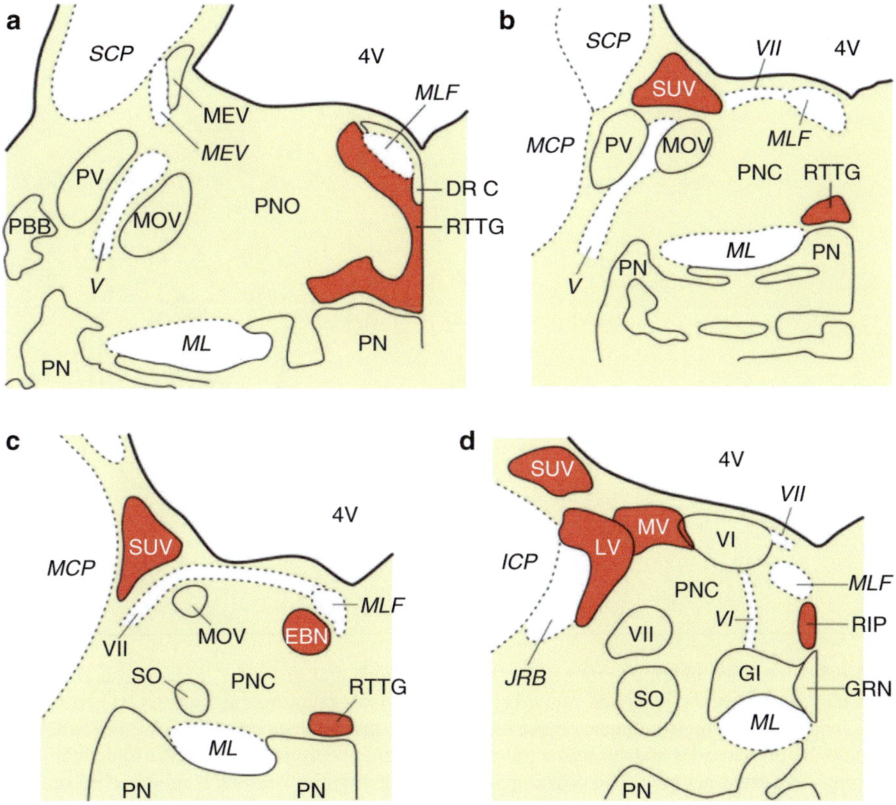

Fig. 6.2 The premotor oculomotor nuclei of the human pons and medulla oblongata. (**a**) Schematized horizontal section through the mid-level of the pons with the reticulotegmental nucleus of the pons (RTTG; nucleus Bechterew) and the medial longitudinal fascicle (*MLF*). (**b**) Schematized horizontal section through the mid-level of the pons, superior vestibular nucleus (*SUV*), caudal portion of the RTTG, and *MLF*. (**c**) Horizontal section through the caudal pons showing the area of the excitatory burst neurons for horizontal saccades (EBN), the caudal nucleus of the pontine reticular formation (*PNC*), the SUV and the most caudal portion of the RTTG. (**d**) The caudal pons with the superior (*SUV*), lateral (*LV*), and medial vestibular nuclei (*MV*), the abducens nucleus (VI) and nerve (*VI*), the raphe interpositus nucleus (*RIP*), and the *MLF*. (**e**) Horizontal section through the pontomedullary junction with the prepositus hypoglossal nucleus (*PPH*), dorsal paragigantocellular reticular nucleus (*DPGI*), MV, and the *MLF* (Modified according to Rüb et al. (2008b), (Figure 2, page 172); with kind permission from Springer Science and Business Media). Abbreviations: *AR* arcuate nucleus, *DC* dorsal cochlear nucleus, *DPGI* dorsal paragigantocellular reticular nucleus, *DR C* dorsal raphe nucleus, caudal part, *EBN* area of the excitatory burst neurons for horizontal saccades, *GI* gigantocellular reticular nucleus, *GRN* great raphe nucleus, *ICP* inferior cerebellar peduncle, *IO* inferior olive, *JRB* juxtarestiform body, *LV* lateral vestibular nucleus, *MCP* medial cerebellar peduncle, MEV mesencephalic trigeminal nucleus, *MEV* mesencephalic trigeminal tract, *ML* medial lemniscus, *MLF* medial longitudinal fascicle, *MOV* motor trigeminal nucleus, MV medial vestibular nucleus, *PBB* pontobulbar body, *PN* pontine nuclei, *PNC* pontine reticular formation, caudal nucleus, *PNO* pontine reticular formation, oral nucleus, *PPH* prepositus hypoglossal nucleus, *PV* principal trigeminal nucleus, *RIP* raphe interpositus nucleus, *RTTG* reticulotegmental nucleus of the pons (nucleus Bechterew), *SCP* superior cerebellar peduncle, SO superior olive, *SUV* superior vestibular nucleus, *SV* spinal vestibular nucleus, *VC* ventral cochlear nucleus, *4V* fourth ventricle, *V* trigeminal nerve, VI abducens nucleus, *VI* abducens nerve, VII facial nucleus, *VII* facial nerve, *VIIIv* vestibulocochlear nerve, vestibular portion

Fig. 6.2 (continued)

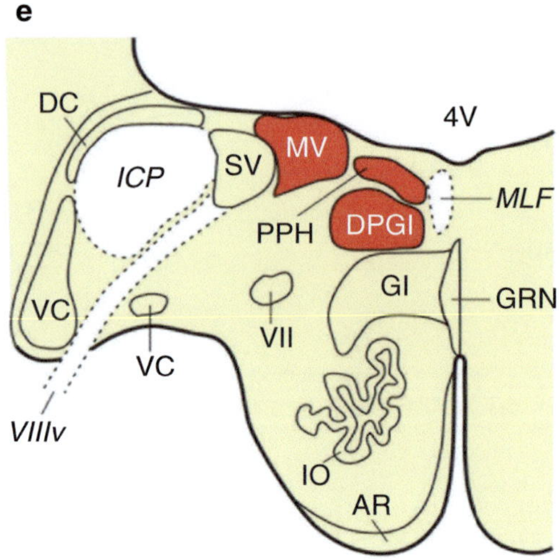

Fig. 6.3 Simplified block diagram of the neural network that generates horizontal and vertical saccades. Saccades are generated by two types of premotor brainstem nerve cells: (1) burst neurons, which deliver high-frequency bursts of activity to the motoneurons of involved external eye muscles before saccades and during saccades, and (2) pause neurons, which exert a tonic inhibition on brainstem burst neurons. The ongoing activity of these pause neurons is inhibited before and during saccades in a specific direction (directional pause neurons) or in all directions (omnipause neurons). In humans, the excitatory burst neurons for horizontal saccades are located in a subnucleus of the caudal portion of the pontine reticular formation (i.e., area of the excitatory burst neurons for horizontal saccades, *EBN*) and the excitatory burst neurons for vertical saccades in the midbrain rostral interstitial nucleus of the medial longitudinal fascicle (*riMLF*). The inhibitory burst neurons for horizontal saccades reside in the dorsal paragigantocellular reticular nucleus (*DPGI*). These inhibitory DPGI nerve cells are activated by the burst neurons of the EBN and inhibit the contralateral abducens nucleus (*VI*) during horizontal saccades. Located in the pontomedullary raphe interpositus nucleus (*RIP*), the saccadic omnipause neurons fire with a high ongoing rate before and during saccades. The pause of the ongoing firing of these omnipause neurons in the RIP allows the burst neurons in the EBN and riMLF to activate the external eye muscle motoneurons in the oculomotor cranial nerve nuclei (i.e., oculomotor, trochlear, and abducens nuclei, III, IV, VI). The omnipause neurons in the RIP send inhibitory projections to the premotor burst neurons in the EBN and riMLF and excitatory efferents to the inhibitory burst neurons of the DPGI. While the EBN, riMLF, and DPGI receive excitatory projections from the superior colliculus (*SC*), the efferent projections from the SC to the saccadic omnipause neurons in the RIP inhibit their tonic firing activity before and during saccades in all directions. Projections from the frontal and supplementary eye fields supply the omnipause neurons in the RIP with cortical information. The subcortical sites that until now have been shown to sustain neuronal loss in Huntington's disease (HD) are depicted in white (Modified according to Rüb et al. (2003a), (Figure 1, page 117); with kind permission from Elsevier). Abbreviations: *DPGI* dorsal paragigantocellular reticular nucleus, *EBN* area of the excitatory burst neurons for horizontal saccades, *riMLF* rostral interstitial nucleus of the medical longitudinal fascicle, *RIP* raphe interpositus nucleus, *III* oculomotor nucleus, *IV* trochlear nucleus, *VI* abducens nucleus

and lateral vestibular (LV) nuclei, the area of the excitatory burst neurons for horizontal saccades (EBN), the raphe interpositus nucleus (RIP), and the medial vestibular (MV), the prepositus hypoglossal (PPH), and the dorsal paragigantocellular reticular (DPGI) nuclei (Figs. 6.1, 6.2, 6.3, 6.4, and 6.5) (Büttner-Ennever 2006; Büttner and Büttner-Ennever 2006; Büttner-Ennever and Gerrits 2004; Büttner-Ennever and Horn 1997, 2004; Büttner-Ennever et al. 1982; Horn 2006; Horn et al.

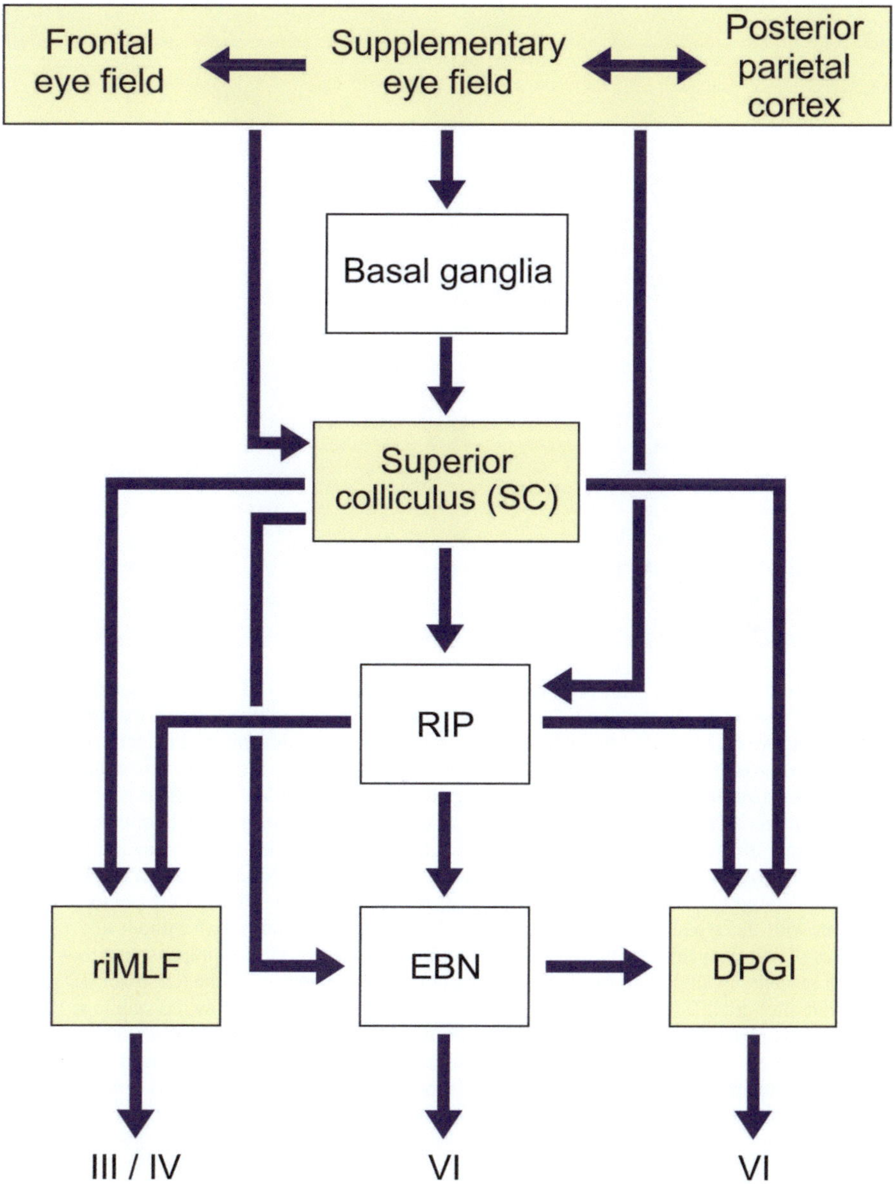

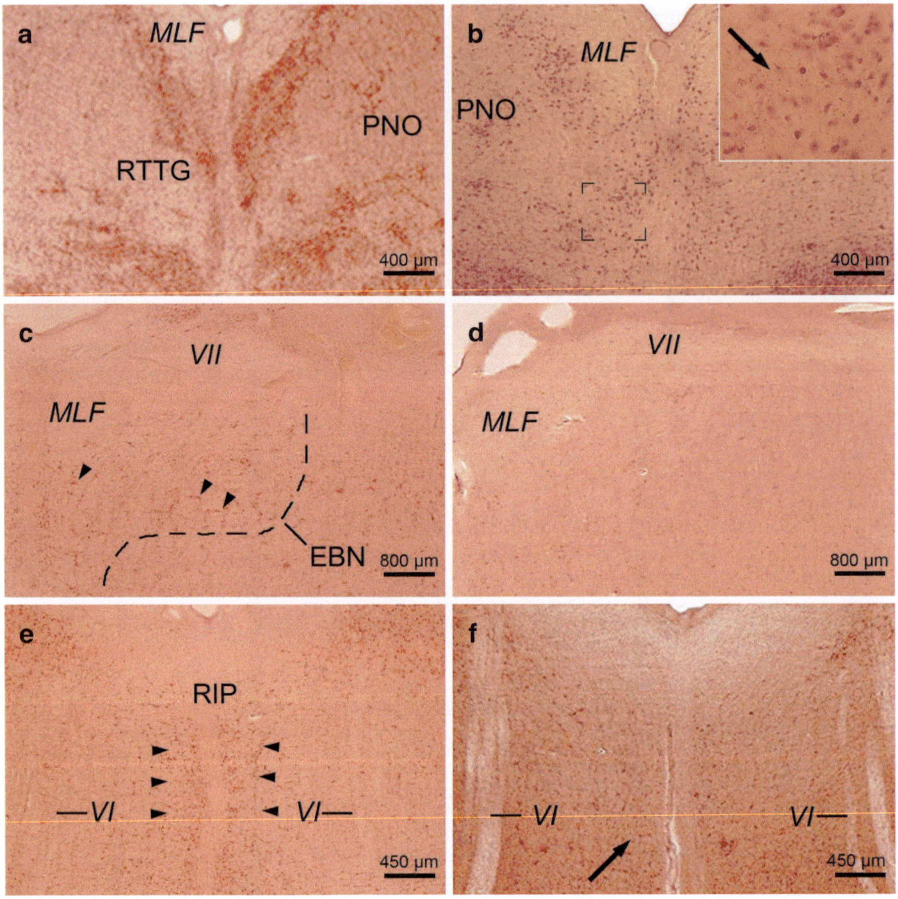

Fig. 6.4 Neuronal loss in mid-pontine and pontomedullary nuclei of the premotor oculomotor brainstem network in Huntington's disease (HD). (**a**) Horizontal section through the mid-pons of a representative control individual depicting the oculomotor reticulotegmental nucleus of the pons (*RTTG*). (**b**) Loss of RTTG nerve cells in a typical Huntington's disease (HD) patient with Vonsattel grade 2 of neostriatal atrophy. *Inset*: remaining RTTG nerve cells and lipofuscin deposits in the neuropil (*arrow*). (**c**) Horizontal section through the caudal pons of a typical control individual with the area of the excitatory burst neurons for horizontal saccades (*EBN*) showing its typical horizontally oriented nerve cells (*arrowheads*). (**d**) Absence of EBN nerve cells of an HD patient with Vonsattel grade 4 of neostriatal atrophy. (**e**) Horizontal section through the pontomedullary junction with the raphe interpositus nucleus (*RIP*) of a control individual without any medical histories of neuropsychiatric diseases. *Arrowheads* point to the typical omnipause neurons of the RIP for saccades in all directions. (**f**) Severe loss of omnipause neurons in the RIP of an HD patient with Vonsattel grade 2 of neostriatal atrophy. (**a–f** aldehyde-fuchsin Darrow red staining; 100 µm polyethylene glycol sections) (Reprinted from Rüb et al. (2014a), (Figure 2, page 250); with kind permission from John Wiley and Sons). Abbreviations: *MLF* medial longitudinal fascicle, *PNO* pontine reticular nucleus, oral subnucleus, *RIP* raphe interpositus nucleus, *RTTG* reticulotegmental nucleus of the pons (nucleus Bechterew), *VI* abducens nerve, *VII* facial nerve

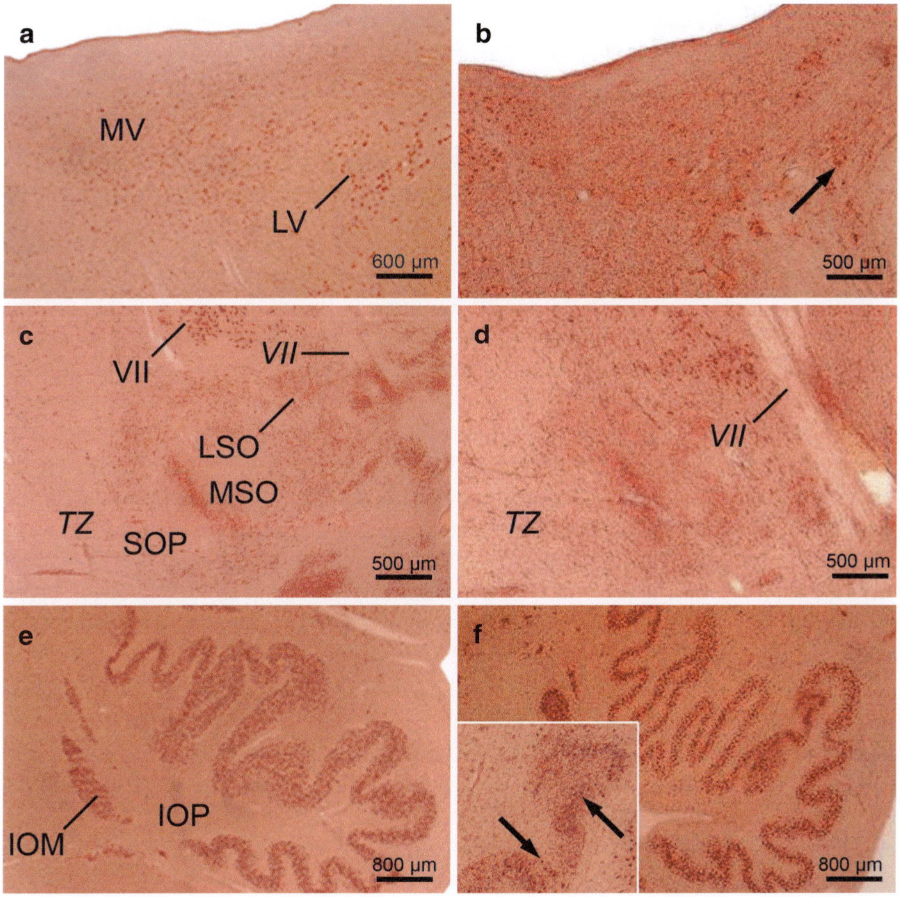

Fig. 6.5 Neuronal loss in the pontomedullary junction and medulla oblongata in Huntington's disease (HD). (**a**) Horizontal section through the pontomedullary junction of a typical control individual with the lateral vestibular (*LV*) and magnocellular portion of the medial vestibular (*MV*) nuclei. (**b**) Severe neuronal loss in the LV and MV of a Huntington's disease (HD) patient with Vonsattel grade 3 of neostriatal atrophy. *Arrow*: remaining giant LV nerve cell. (**c**) Horizontal section through the mid-pons of a representative control individual depicting the auditory superior olive: (*MSO*) medial and (*LSO*) lateral subnuclei, (*SOP*) periolivary group. (**d**) Neuronal loss in the superior olive of an HD patient with Vonsattel grade 3 of neostriatal atrophy predominantly affecting the LSO and SOP. (**e**) Horizontal section through the medulla oblongata of a control individual without any medical histories of neuropsychiatric diseases showing the inferior olive with its principal (*IOP*) and medial subnuclei (*IOM*). (**f**) Severe neuronal loss in circumscribed regions of the IOP (*asterisk*) of an HD patient with Vonsattel grade 2 of neostriatal atrophy. *Inset*: devastated area of the IOP with severe neuronal loss and the presence of extraneuronal lipofuscin deposits (*arrows*: extraneuronal lipofuscin deposits which can serve as reliable marker for neurodegeneration). (**a–f** aldehyde-fuchsin Darrow red staining; 100 µm polyethylene glycol sections) (Reprinted from Rüb et al. (2014a), (Figure 3, page 251; with kind permission from John Wiley and Sons). Abbreviations: *IOM* inferior olive, medial subnucleus, *IOP* inferior olive, principal subnucleus, *LSO* lateral superior olive, *LV* lateral vestibular nucleus, *MSO* medial superior olive, *MV* medial vestibular nucleus, magnocellular portion, *SOP* Superior olive, periolivary group, *TZ* trapezoid body, *VII* Facial nerve

1994, 1995, 1999; Leigh and Kennard 2004; Leigh and Zee 2006; May 2006; Rüb et al. 2001, 2003a, 2004a, b, 2008b, 2009).

6.4 The Human Premotor Oculomotor Brainstem Nuclei in Pigment-Nissl Stained Thick Serial Tissue Sections

The human *riMLF* represents the immediate premotor structure for vertical and torsional saccades and lies in the rostral midbrain between the thalamus and the red nucleus. It is dorsally bordered by the thalamo-subthalamic paramedian artery, appears as a wing-shaped and reticular nucleus in close proximity to the subparafascicular nucleus of the thalamic central complex, and is separated from the immediately caudally located C by the tractus retroflexus (habenulo-interpeduncular tract). The elongated nerve cells of the human riMLF are loosely intermingled between the fiber bundles of the *MLF* and represent the premotor burst neurons necessary for the generation of vertical and torsional saccades (Figs. 6.1 and 6.3) (Büttner-Ennever 2006; Büttner and Büttner-Ennever 2006; Büttner-Ennever and Gerrits 2004; Büttner-Ennever and Horn 1997, 2004; Büttner-Ennever et al. 1982; Horn 2006; Horn et al. 1994, 1995, 1999; Leigh and Kennard 2004; Leigh and Zee 2006; May 2006; Rüb et al. 2001, 2003a, 2004a, b, 2008b, 2009).

The adjacent *C* represents a small and compact nucleus of the rostral midbrain, which is embedded into the *MLF* and is dorsolaterally bordered by the nucleus of Darkschewitsch. The human C is considered the major structure for vertical and torsional gaze holding (Fig. 6.1) (Büttner-Ennever 2006; Büttner and Büttner-Ennever 2006; Büttner-Ennever and Horn 2004; Horn 2006; Horn et al. 1999; Leigh and Kennard 2004; Leigh and Zee 2006; Rüb et al. 2001, 2008b, 2009).

The human *SC* represents a layered structure, forms a mound-like elevation of the rostral half of the midbrain tectal plate, and is crucial for the orienting response to visual or auditory stimuli. It is involved in the initiation of saccades, contributes to vergence, and is important for the suppression of unwanted saccades when steady fixation of a visual target is necessary (Figs. 6.1 and 6.3) (Büttner and Büttner-Ennever 2006; Büttner-Ennever and Horn 1997; Horn 2006; Leigh and Zee 2006; May 2006; Rüb et al. 2008b, 2009).

The human *RTTG* (*nucleus Bechterew*) with its two processes is located close to the midline of the pons and extends from the central raphe nucleus rostrally to the rostral pole of the great raphe nucleus caudally. The dorsolateral process of the human RTTG abuts on the *MLF* and the ventrolateral one on the medial lemniscus. The human RTTG is a crucial component of the brain neural circuit engaged in the regulation and control of smooth pursuit eye movements, guarantees the accuracy of horizontal saccades, and harbors vergence-related nerve cells (Fig. 6.2) (Bütttner and Büttner-Ennever 2006; Büttner-Ennever and Horn 1997; Horn 2006; Rüb et al. 2004b, 2008b, 2009).

The human *SUV* is situated on the floor of the fourth ventricle in the caudal portion of the pons, extends from the rostral pole of the motor and principal trigeminal nuclei up to the caudal pole of the superior olive, and is integrated into the neural circuits that generate the vestibulo-ocular reflex and the optokinetic nystagmus (Figs. 5.7 and 6.2) (Büttner and Büttner-Ennever 2006; Büttner-Ennever and Gerrits 2004; Leigh and Zee 2006; Rüb et al. 2004a 2008b, 2009).

The *LV* lies at the pontomedullary junction, is embedded into the juxtarestiform body, is architectonically characterized by the presence of quadrangular or multipolar and strongly pigmented giant neurons (i.e., Deiter's neurons), and contributes to the performance of the optokinetic nystagmus (Figs. 5.7, 6.2, and 6.5) (Leigh and Zee 2006; Rüb et al. 2004a, 2008b, 2009).

The *EBN* and the *RIP* are located in the *paramedian pontine reticular formation* (*PPRF*), which occupies the major portion of the pontine tegmentum and consists of an oral (PNO) and a caudal nucleus (PNC). The PNC contains two distinct areas, which play an important role in the generation of saccades, the area of the excitatory burst neurons for horizontal saccades and the area of the omnipause neurons (Figs. 6.2, 6.3, and 6.4).The human EBN lies ventrolaterally to the *MLF* and immediately underneath the facial nerve, immediately rostral to the abducens nucleus and the RIP, and is only about 2.5 mm in width and 2 mm in height. The human EBN is crucial for the generation of horizontal saccades and provides the immediate premotor signals for these rapid horizontal eye movements (Figs. 6.2, 6.3, and 6.4). The very small human RIP is confined to the pontine level at which the rootlets of the abducens nerve pass through the brainstem. Its strikingly linearly arranged nerve cells are oriented with their long axis perpendicular to the midline, form a narrow band of only two or three nerve cells on either side of the midline, and overwhelmingly represent saccadic omnipause neurons, which act as a trigger for the initiation of saccades in all directions (Figs. 6.2, 6.3, and 6.4) (Büttner-Ennever 2006; Büttner and Büttner-Ennever 2006; Büttner-Ennever and Horn 2004; Horn 2006; Horn et al. 1994, 1995; Leigh and Kennard 2004; Leigh and Zee 2006; Rüb et al. 2003a, 2008b, 2009).

The human *MV* lies at the pontomedullary junction and is composed of a laterally situated magnocellular as well as a medially located parvocellular part. The MV subserves the vestibulo-ocular reflex and together with the PPH participates in horizontal gaze holding (Figs. 5.7, 6.2, and 6.5) (Büttner-Ennever 2006; Büttner and Büttner-Ennever 2006; Büttner-Ennever and Horn 2004; Horn 2006; Horn et al. 1994, 1995; Leigh and Kennard 2004; Leigh and Zee 2006; Rüb et al. 2003a, 2008b, 2009).

The human *PPH* is located on the floor of the fourth ventricle in the pontomedullary junction and between the oral pole of the hypoglossal nucleus and the caudal pole of the abducens nucleus and, together with the MV, is crucial for horizontal gaze holding (Fig. 6.2) (Büttner-Ennever 2006; Büttner and Büttner-Ennever 2006; Rüb et al. 2008b, 2009).

The *DPGI* is located in the dorsomedial part of the pontomedullary tegmentum and lies medial to the *MLF*, ventral to the PPH, and dorsal to the gigantocellular reticular nucleus. Its loosely and irregularly arranged nerve cells represent inhibitory burst neurons for vertical saccades (Figs. 6.2 and 6.3) (Büttner-Ennever 2006; Büttner and Büttner-Ennever 2006; Büttner-Ennever and Horn 2004; Horn 2006; Horn et al. 1994, 1995; Leigh and Kennard 2004; Leigh and Zee 2006; Rüb et al. 2003a, 2008b, 2009).

6.5 In Search for the Morphological Correlates of Oculomotor Dysfunctions in Huntington's Disease (HD)

Until a few years ago, little was known about the *morphological correlates of oculomotor dysfunctions in HD*. Pathophysiological explanations of the oculomotor dysfunctions of HD patients have most frequently focused exclusively on the most prominent oculomotor abnormalities (i.e., delay and slowing of saccades, fixation instability) and as practiced with other HD symptoms over many decades, have been mostly attributed to the well-known *interruption of the basal ganglia loops* (Blekher et al. 2004, 2006; Bollen et al. 1986; Collewijn et al. 1988; Lasker and Zee 1997; Leigh and Zee 2006; Leigh et al. 1983; Mink 1996; Oepen et al. 1981, 1985; Ortega et al. 2007; Rüb et al. 2009; Tian et al. 1991; Walker 2007a; Winograd-Gurvich et al. 2003). Nevertheless, theoretical attempts have been made to explain a hypothesized *contribution of brainstem lesions* to the pathogenesis of some oculomotor HD dysfunctions (i.e., damage to neural circuits, including the SC attributed to initiating deficits of saccades; degeneration of the saccadic burst neurons in the EBN and/or in the riMLF or affection of the omnipause neurons in the RIP attributed to saccadic slowing, degeneration of the omnipause neurons of the RIP or nerve cells of the SC attributed to dysmetrical saccades, affection of neural circuits, including the SC to fixation instability) (Blekher et al. 2004, 2006; Bollen et al. 1986; Lasker and Zee 1997; Leigh and Zee 2006; Leigh et al. 1983; Oepen et al. 1981, 1985; Rüb et al. 2009; Tian et al. 1991; Winograd-Gurvich et al. 2003). Until recently the occurrence of oculomotor dysfunctions could not be unequivocally linked to the affection of the premotor oculomotor brainstem nuclei in HD patients (Kennard and Lueck 1989; Koeppen 1989; Leigh et al. 1985; Rüb et al. 2009), because the premotor oculomotor brainstem nuclei (1) have not been considered in previous neuropathological HD studies (i.e., SC, RTTG, MV, SUV), (2) could not be exactly delineated anatomically due to inaccurate previous neuroanatomical data (i.e., EBN, RIP), or (3) were investigated only in thin single thin brain tissue sections normally used for diagnostic neuropathological investigations (i.e., riMLF) (Table 6.2).

Usage of the improved knowledge in the field of the functional neuroanatomy of the human premotor oculomotor brainstem network in *pathoanatomical investigations* of unconventional 100 μm-thick serial tissue sections stained for neuronal Nissl substance and lipofuscin pigment (i.e., *pigment-Nissl stain according to Braak*) has been suggested as a promising strategy (1) to systematically study the premotor oculomotor brainstem network in HD, (2) to elucidate the neuroanatomical basis of the large spectrum of oculomotor dysfunctions in HD patients, and (3) to clarify the contribution of the nuclei of the *human premotor oculomotor brainstem network* to the pathogenesis of oculomotor dysfunctions in HD (Braak et al. 2003b;

Table 6.2 The human premotor oculomotor brainstem nuclei: functional consequences of their damage and previous findings in Huntington's disease (HD)

Premotor brainstem nucleus	Functional consequences of lesions	Previous pathoanatomical findings in Huntington's disease (HD)
Rostral interstitial nucleus of the medial longitudinal fascicle (riMLF)	Slowing of vertical and torsional saccades; loss or restricted range of vertical and torsional saccades	No clear evidence for degeneration or intactness in HD
Colliculus superior (SC)	Increased saccade latencies, saccadic intrusions	Not investigated in HD
Reticulotegmental nucleus of the pons (RTTG)	Saccadic smooth pursuits, dysmetrical saccades	Not investigated in HD
Superior vestibular nucleus (SUV)	Reduced gain or loss of vestibulo-ocular reflex and optokinetic nystagmus	Not investigated in HD
Lateral vestibular nucleus (LV)	Reduced gain or loss of optokinetic nystagmus	Degeneration suggested in HD
Area of the excitatory burst neurons for horizontal saccades (EBN)	Slowing of horizontal and torsional saccades; loss or restricted range of horizontal and torsional saccades	Unclear – inappropriately Delineated in HD
Raphe interpositus nucleus (RIP)	Slowed vertical and horizontal saccades	Unclear – inappropriately delineated in HD
Medial vestibular nucleus (MV)	Reduced gain or loss of vestibulo-ocular reflex, horizontal gaze-evoked nystagmus	Not investigated in HD

Modified according to Rüb et al. (2009), (Table 2, page 711); with kind permission from John Wiley and Sons

Abbreviations: *EBN* area of the excitatory burst neurons for horizontal saccades, *LV* lateral vestibular nucleus, *MV* medial vestibular nucleus, *riMLF* rostral interstitial nucleus of the medial longitudinal fascicle, *RIP* raphe interpositus nucleus, *RTTG* reticulotegmental nucleus of the pons, *SC* superior colliculus, *SUV* superior vestibular nucleus

Rüb et al. 2008b, 2009). Investigations of unconventional 100 μm-thick serial tissue sections stained according to the pigment-Nissl method (1) are complementary to the *conventional diagnostic neuropathological procedure* (i.e., investigation of single, thin brain tissue sections), (2) allow for unequivocal identification of all known discrete nuclei of the premotor oculomotor brainstem network and help to study their possible involvement in neurodegenerative diseases, (3) have revealed improved insights into the morphological basis of a large spectrum of oculomotor dysfunctions in different neurodegenerative diseases (Braak et al. 2003b; Büttner-Ennever 2006; Büttner and Büttner-Ennever 2006; Büttner-Ennever et al. 1982; Büttner-Ennever and Horn 1997, 2004; Horn 2006; Horn et al. 1994, 1995, 1999; Leigh and Kennard 2004; Leigh and Zee 2006; Rüb et al. 2001, 2003a, 2004a, b, 2008b, 2009; Seidel et al. 2015), and (4) have therefore recently been applied for the first time in HD brainstem research (Figs. 6.4 and 6.5) (Rüb et al. 2014a).

6.6 Degeneration of Select Nuclei of the Human Premotor Oculomotor Brainstem Network in Huntington's Disease (HD) and its Clinical Significance

In view of the close parallels between the pathological features of the various oculomotor dysfunctions in HD (Table 6.2) and the resulting main effects of damage to the nuclei of the premotor oculomotor brainstem network in humans and nonhuman primates, it appeared likely that some of these nuclei are among the targets of the neurodegenerative process of HD (Rüb et al. 2008b, 2009). Indeed, systematic pathoanatomical studies performed on unconventional thick tissue sections stained for neuronal Nissl material and lipofuscin pigment revealed a *consistent neuronal loss in the RTTG, EBN, RIP, SUV, MV, and LV* of HD patients (Figs. 4.3, 6.3, 6.4, and 6.5).

As a result, these new findings in premotor oculomotor nuclei of the pons, pontomedullary junction, and medulla oblongata for the first time offered adequate explanations for a variety of oculomotor dysfunctions that may occur in HD patients (Rüb et al. 2014a). Considering its crucial role in the circuits engaged in the regulation and control of smooth pursuit eye movements, damage to the *RTTG* together with the lesions in the interconnected fastigial nucleus of the cerebellum is well suited to explain why slowed and saccadic smooth pursuits occur in HD patients (Table 6.2) (Rüb et al. 2008b, 2009, 2014a). As the loss of the premotor omnipause neurons in the *RIP* has recently been shown to correlate with slowing of saccades, the severe damage to the RIP present in HD patients most likely is responsible for the slowing of their horizontal saccades (Fig. 6.3; Table 6.2) (Rüb et al. 2008b, 2009, 2014a). Since the excitatory burst neurons of the *EBN* are essential for the initiation and generation of horizontal saccades, damage to the EBN of HD patients may account for the initiation deficits and slowing of their horizontal saccades (Fig. 6.3; Table 6.2) (Rüb et al. 2008b, 2009, 2014a). Finally, in view of their known functional role, the involvement of the *MV, LV, and SUV* most likely contributes substantially to the impairments of the optokinetic nystagmus and vestibulo-ocular reaction of HD patients (Fig. 5.7; Table 6.2) (Rüb et al. 2008b, 2009, 2014a).

Although the demonstration of degeneration in the RTTG, RIP, EBN, SUV, LV, and MV revealed improved and new insights into the morphological base of a variety of oculomotor dysfunctions of HD patients, the midbrain nuclei of the premotor oculomotor brainstem network (i.e., *C, SC, riMLF*) still remain to be studied in HD (Rüb et al. 2008b, 2009, 2014a).

This chapter was written by Udo Rüb.

Chapter 7
Widespread Brainstem Neurodegeneration in Huntington's Disease (HD)

7.1 Systematic Pathoanatomical Investigations of the Brainstem in Huntington's Disease (HD)

The involvement of the brainstem is still not among the established degenerative features of HD (Bruyn et al. 1979; Koeppen 1989; Lange and Aulich 1986; Oyanagi et al. 1989; Rüb et al. 2014a; Vonsattel and DiFiglia 1998; Vonsattel et al. 1985). Although some of the unexplained HD disease symptoms (e.g., oculomotor dysfunctions) also suggested degeneration of select brainstem nuclei, damage to the brainstem and its possible clinical relevance in HD has been controversially discussed for many decades (see Sects. 6.2, 6.5, and 6.6). Previous assumptions that select nuclei of the brainstem may also undergo neurodegeneration during HD (i.e., dopaminergic and GABAergic substantia nigra, auditory superior olive, lateral vestibular nucleus, precerebellar inferior olive) (Figs. 1.4, 7.1, 7.2, and 7.3) have not been revisited or taken into account by current researchers (see Chap. 1) (Bruyn et al. 1979; Bollen et al. 1986; Koeppen 1989; Kremer et al. 1992; Lange and Aulich 1986; Lange et al. 1976; Lasker and Zee 1997; Leigh et al. 1983, 1985; Oyanagi et al. 1989; Rüb et al. 2009, 2014a; Vonsattel 2008; Vonsattel and DiFiglia 1998; Vonsattel et al. 1985; Walker 2007a, b). These assumptions, however, were confirmed by recent *systematic pathoanatomical investigations* of serial brainstem sections stained for neuronal Nissl material (with Darrow red) and lipofuscin pigment (with aldehyde-fuchsin) of clinically diagnosed and genetically confirmed HD patients that suffered from chorea, cognitive decline, and personality changes and also showed disease signs possibly related to brainstem damage (i.e., broad-based gait, gait imbalance, dysphagia, slowed horizontal saccades, inability to start horizontal saccades without initial head thrust, slowed and saccadic smooth pursuits, falls) (Rüb et al. 2014a).

U. Rüb et al., *The Neuropathology of Huntington's Disease: Classical Findings, Recent Developments and Correlation to Functional Neuroanatomy*,
Advances in Anatomy, Embryology and Cell Biology, Vol. 217,
DOI 10.1007/978-3-319-19285-7_7

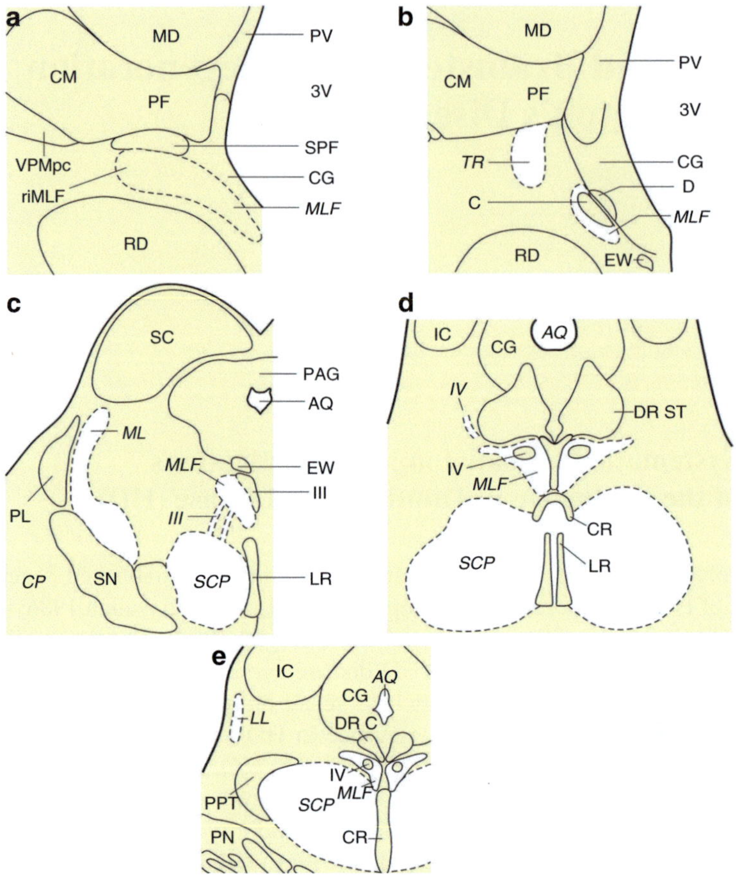

Fig. 7.1 The human midbrain. (**a**) Frontal section through the rostral midbrain at the level of the rostral interstitial nucleus of the medial longitudinal fascicle (*riMLF*). (**b**) Frontal section cut perpendicularly to the intercommissural axis of Forel showing the rostral midbrain at the level of the interstitial nucleus of Cajal (*C*). (**c**) Horizontal section through the caudal midbrain at the level of the dopaminergic substantia nigra (*SN*). (**d**) Horizontal section through the caudal midbrain at the level of the supratrochlear subnucleus of the serotonergic dorsal raphe nucleus (*DR ST*). (**e**) Horizontal section through the caudal midbrain at the level of the cholinergic pedunculopontine nucleus (*PPT*). Abbreviations: *AQ* aqueduct, *CG* central gray, *CM* centromedian nucleus of the thalamus, *CR* central raphe nucleus, *CP* cerebral peduncle, *D* nucleus of Darkschewitsch, *DR C* dorsal raphe nucleus, caudal subnucleus, *DR ST* dorsal raphe nucleus, supratrochlear subnucleus, *EW* Edinger-Westphal nucleus, *C* interstitial nucleus of Cajal, *CR* central raphe nucleus, *IC* Inferior colliculus, *LL* lateral lemniscus, *LR* linear raphe nucleus, *MD* mediodorsal nucleus of the thalamus, *ML* medial lemniscus, *MLF* medial longitudinal fascicle, *PAG* periaqueductal gray, PF parafascicular nucleus of the thalamus, *PL* paralemniscal nucleus, *PN* pontine nuclei, *PV* paraventricular nucleus of the thalamus, *R* red nucleus, *riMLF* rostral interstitial nucleus of the medial longitudinal fascicle, *SC* superior colliculus, *SCP* superior cerebellar peduncle, *SN* substantia nigra, *SPF* subparafascicular nucleus of the thalamus, *TR* tractus retroflexus (habenulo-interpeduncular tract), *VPM* ventral posterior medial nucleus of the thalamus, *VPMpc* ventral posterior medial nucleus of the thalamus, parvocellular part, III oculomotor nucleus, *III* oculomotor nerve, *3V* third ventricle, IV trochlear nucleus, *IV* trochlear nerve

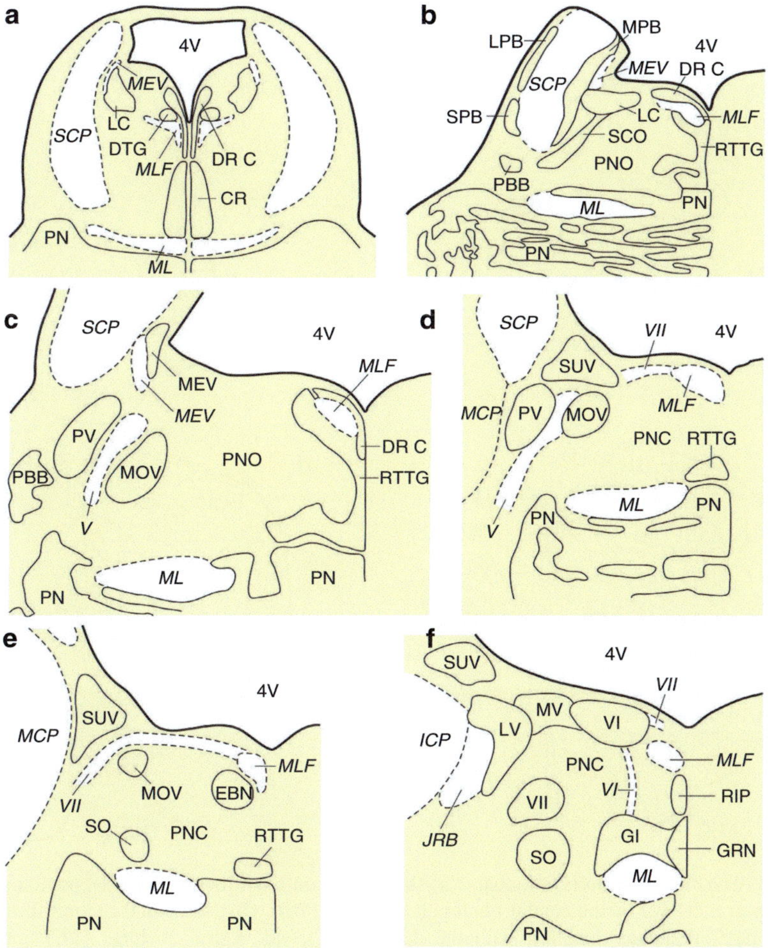

Fig. 7.2 The human pons. (**a**) Horizontal section through the rostral pons at the level of the central raphe nucleus (*CR*). (**b**) Horizontal section through the rostral pons at the level of the parabrachial nuclei (i.e., *MPB* medial parabrachial nucleus, *LPB* lateral parabrachial nucleus, *SPB* subpeduncular pigmented nucleus). (**c**) Horizontal section through the mid pons at the level of the trigeminal nuclei (i.e., *MOV* motor trigeminal nucleus, *PV* principal trigeminal nucleus, *MEV* mesencephalic trigeminal nucleus). (**d**) Horizontal section through the caudal pons at the level of the superior vestibular nucleus (*SUV*). (**e**) Horizontal section through the caudal pons at the level of the area of the excitatory burst neurons for horizontal saccades (*EBN*). (**f**) Horizontal section through the pontomedullary junction at the level of the raphe interpositus nucleus (*RIP*). Abbreviations: *CR* central raphe nucleus, *DR C* dorsal raphe nucleus, caudal part, *DTG* dorsal tegmental nucleus (Gudden), *EBN* area of the excitatory burst neurons for horizontal saccades, *GI* gigantocellular reticular nucleus, *GRN* great raphe nucleus, *ICP* inferior cerebellar peduncle, *JRB* juxtarestiform body, *LC* locus coeruleus, *LPB* lateral parabrachial nucleus, *LV* lateral vestibular nucleus, *MCP* medial cerebellar peduncle, *MEV* mesencephalic trigeminal nucleus, *MEV* mesencephalic trigeminal tract, *ML* medial lemniscus, *MLF* medial longitudinal fascicle, *MPB* medial parabrachial nucleus, *MOV* motor trigeminal nucleus, *MV* medial vestibular nucleus, *PBB* pontobulbar body, *PN* pontine nuclei, *PNC* pontine reticular formation, caudal nucleus, PNO pontine reticular formation, oral nucleus, *PV* principal trigeminal nucleus, *RIP* raphe interpositus nucleus, *RTTG* reticulotegmental nucleus of the pons (nucleus Bechterew), *SCO* subcoeruleus nucleus, *SCP* superior cerebellar peduncle, *SO* superior olive, *SPB* subpeduncular pigmented nucleus, *SUV* superior vestibular nucleus, *4V* fourth ventricle, *V* trigeminal nerve, VI abducens nucleus, *VI* abducens nerve, VII facial nucleus, *VII* facial nerve

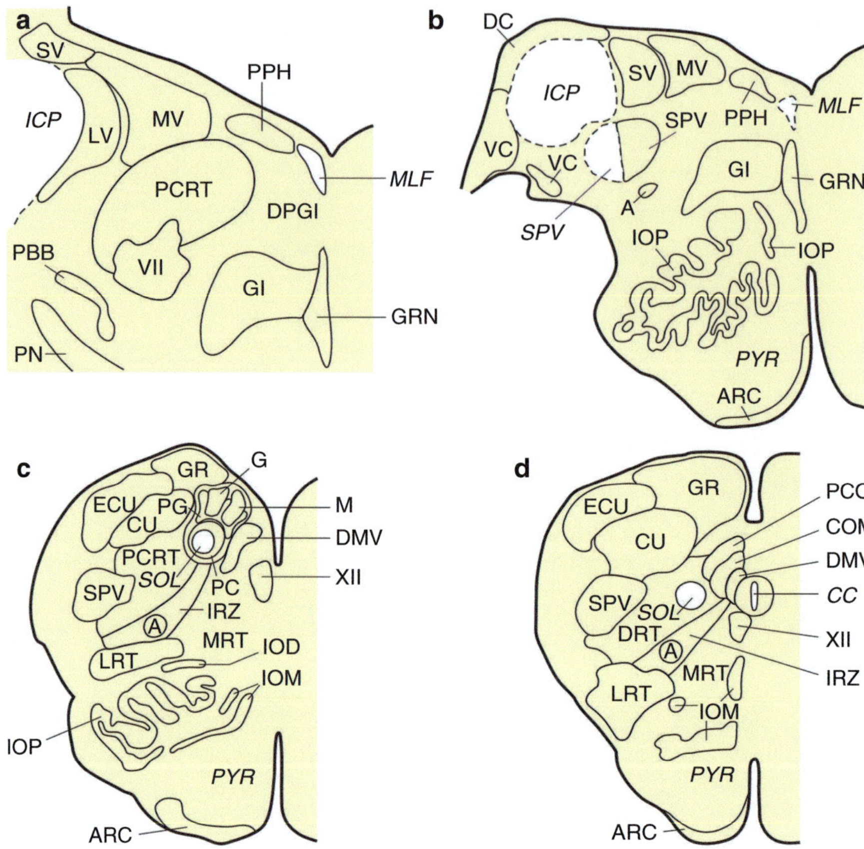

Fig. 7.3 The human medulla oblongata. (**a**) Horizontal section through the caudal pontomedullary junction at the level of the caudal pontobulbar body (*PBB*). (**b**) Horizontal section through the caudal pontomedullary junction with the dorsal (*DC*) and ventral cochlear (*VC*) nuclei. (**c**) Horizontal section through the rostral medulla oblongata with the principal nucleus of the inferior olive (*IOP*). (**d**) Horizontal section through the caudal medulla oblongata with the central canal (*CC*). Abbreviations: *A* ambiguus nucleus, *ARC* arcuate nucleus, *COM* commissural solitary nucleus, *CU* cuneate nucleus, *DC* dorsal cochlear nucleus, *DMV* dorsal motor vagal nucleus, *DPGI* dorsal paragigantocellular reticular nucleus, *DRT* dorsal reticular nucleus, *ECU* external cuneate nucleus, *G* gelatinous solitary nucleus, *GR* gracile nucleus, *GRN* great raphe nucleus, *ICP* inferior cerebellar peduncle, *IOD* inferior olive, dorsal subnucleus, *IOM* inferior olive, medial subnucleus, *IOP* inferior olive, principal subnucleus, *IRZ* intermediate reticular zone, *LRT* lateral reticular nucleus, *LV* lateral vestibular nucleus, *M* medial solitary nucleus, *MLF* medial longitudinal fascicle, *MRT* medial reticular nucleus, *MV* medial vestibular nucleus, *PBB* pontobulbar body, *PC* parvocellular solitary nucleus, *PCOM* paracommissural solitary nucleus, *PCRT* parvocellular reticular nucleus, *PG* pigmented solitary nucleus, *PN* pontine nuclei, *PPH* prepositus hypoglossal nucleus, *PYR* pyramidal tract, *SOL* solitary tract, *SPV* spinal trigeminal nucleus, *SPV* spinal trigeminal tract, *SV* spinal vestibular nucleus, *VC* ventral cochlear nucleus, *4V* fourth ventricle, VII facial nucleus, *XII* hypoglossal nucleus

The pathoanatomical approach with unconventional thick serial brain tissue sections complements the routine neuropathological examination of single thin brain tissue sections and allows the recognition of subtle pathological changes and reveals more detailed information with respect to the extent and severity of the pathological processes of neurodegenerative diseases than the routine neuropathological approach (Figs. 5.2, 5.3, 5.4, 5.5, 6.4, and 6.5). This pathoanatomical approach was applied in recent systematic HD brainstem studies and disclosed a consistent and widespread degeneration of a variety of brainstem nuclei in HD patients (Rüb et al. 2008b, 2009, 2014a) (Figs. 6.4 and 6.5). Within the framework of these systematic studies, serial thick brainstem sections were also immunolabeled with an antibody directed against the *protein aggregation marker p62* to identify neuronal aggregations of the disease protein of HD. The degradation-related, multifunctional, proteasomal shuttle and ubiquitin-interacting protein p62 together with *ubiquitin* participates in the intracellular trafficking, removal, and degradation of short-lived or aberrantly structured proteins. It serves the presentation of polyubiquitinated protein substrates to the 26S proteasome and interacts with polyubiquitin-tagged protein substrates, as well as with the proteolytic proteasome. The ubiquitin-association domain (UBA) of p62 binds covalently to polyubiquitin chains and enables low-affinity interaction with them. The ubiquitin-like domain (UbL) in the N-terminus of p62 binds to and directly interacts with the multimeric 26S proteasome. p62 is also known to be a structural component of the abnormal nuclear and/or cytoplasmic neuronal aggregations of the polyubiquitinated disease proteins in various human neurodegenerative diseases (see Sects. 8.2 and 8.3) (Figs. 8.3, 8.4, and 8.5) (Kuusisto et al. 2001, 2008; Ortega et al. 2007; Rüb et al. 2014a; Seibenhener et al. 2004; Seidel et al. 2009, 2010; Wooten et al. 2006).

Double immunostaining procedures were performed to identify and examine the topographical and intra-axonal localization, as well as composition of p62 immunoreactive aggregates in the brainstem neuropil. The AT270 antibody, directed against the neuronal cytoskeletal protein tau, was used as axonal marker, and the anti-p62 antibody served as marker for axonal protein aggregates (Figs. 9.2, 9.3, 9.4, and 9.5) (Goedert et al. 1994; Rüb et al. 2014a; Shaw et al. 1981). In addition, the 1C2 antibody was used to confirm the presence of *huntingtin* (*Htt*) with an expanded polyglutamine sequence of more than 37 repeats in the axonal aggregates. An anti-ubiquitin antibody was applied to investigate the additional association of axonal aggregates with the small degradation-related 8 kDa protein ubiquitin assigned to the family of the heat shock proteins (HSP) (Figs. 9.4 and 9.5) (Rüb et al. 2014a; Seidel et al. 2010; Trottier et al. 1995). The immunocytochemical findings of these recent systematic brainstem studies are detailed in Sect. 9.2.

7.2 Neurodegenerative Brainstem Features in Huntington's Disease (HD)

The brainstem of HD patients (Vonsattel grades 2–4 of striatal atrophy) showed no *macroscopical signs of degeneration* except for a slightly flattened pons (Fig. 7.4) (Rüb et al. 2014a). Light microscopical examination of thick brainstem tissue sections, however, showed that brainstem affection in HD is more widespread than believed so far. It revealed a consistent *neuronal loss* in the dopaminergic compact part and GABAergic reticulate part of the *substantia nigra*, in the precerebellar *pontine nuclei* and *inferior olive*, as well as in the oculomotor *reticulotegmental nucleus of the pons*, in the premotor oculomotor *area of the excitatory burst neurons for horizontal saccades* and *raphe interpositus nucleus*, in the auditory *superior olive*, as well as in the *lateral and medial vestibular nuclei*. In some of the HD patients studied, a mild neuronal loss was also present in the superior vestibular nucleus, as well as in the ingestion-related facial and parvocellular reticular nuclei (Figs. 5.7, 6.2, 6.3, 6.4, 6.5, 7.2, and 7.3).

Neuronal loss in the brainstem nuclei did not correlate with the length of the CAG repeats in the mutated HD allele, age at HD onset, duration of HD, age at death, or the Vonsattel grade of striatal atrophy. *GFAP immunoreactive astrocytes* were present in all degenerated and spared brainstem nuclei of all HD patients (Rüb et al. 2014a).

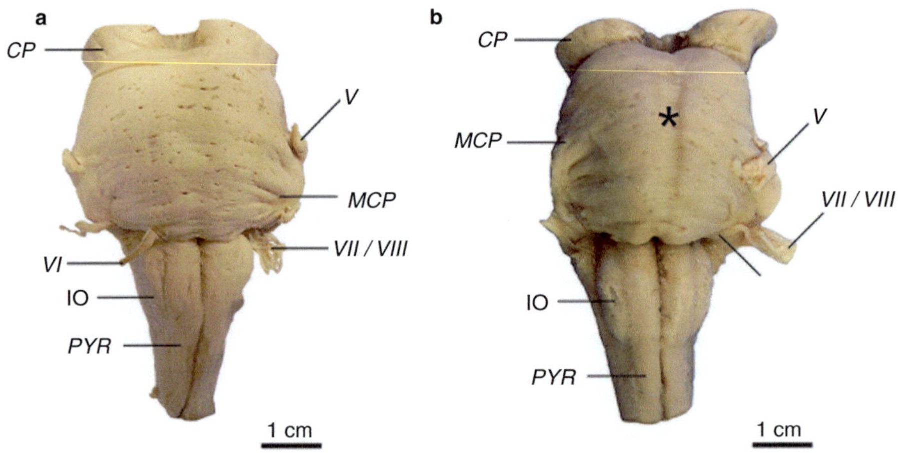

Fig. 7.4 The lower brainstem in Huntington's disease (HD). (**a**) Ventral aspect of the lower brainstem of a typical control individual. (**b**) The lower brainstem with a flattened pons (*asterisk*) of a clinically diagnosed and genetically confirmed Huntington's disease (HD) patient (Vonsattel grade 3 of neostriatal atrophy) (Reprinted from Rüb et al. (2014a), (Figure 1, page 251); with kind permission from John Wiley and Sons). Abbreviations: *CP* cerebral peduncle, *IO* inferior olive, *MCP* medial cerebellar peduncle, *PYR* pyramidal tract, *V* trigeminal nerve, *VI* abducens nerve, *VII* facial nerve, *VIII* vestibulocochlear nerve

7.3 Clinical Relevance of Brainstem Neurodegeneration in Huntington's Disease (HD)

Although its involvement is not yet among the currently established degenerative features of HD, damage to the brainstem underpins that neurodegeneration during HD is not restricted to the neostriatum and select areas of the cerebral neo- and allocortex. It provides additional support for the currently favored opinion that HD is a *multisystem neurodegenerative disease*, which is characterized by a distribution pattern of brain damage that shows more overlap with that of the *related polyglutamine ataxias* (e.g., spinocerebellar ataxias types 1, 2, and 3) than has been believed so far (Atkin and Paulson 2014; Borrell-Pagès et al. 2006; Braak and Braak 1992a, b; Bruyn et al. 1979; De la Monte et al. 1988; Estrada-Sanchez and Rebec 2013; Fennema-Notestine et al. 2004; Ferrante et al. 1987; Finkbeiner and Mitra 2008; Hedreen et al. 1991; Heinsen et al. 1992, 1994, 1996, 1999; Heinsen and Rüb 1997; Imarisio et al. 2008; Lange 1981; Lange and Aulich 1986; Lange et al. 1976; Lastres-Becker et al. 2008; Li and Conforti 2013; Margolis and Ross 2003; Myers et al. 1988; Riess et al. 2008; Rüb et al. 2003a, 2004a, b, 2005, 2008a, 2009, 2013b; Schulte and Littleton 2011; Seidel et al. 2012; Selemon et al. 2004; Sieradzan and Mann 2001; Sotrel et al. 1991; Valera et al. 2005; Vonsattel 2008; Vonsattel and DiFiglia 1998; Vonsattel et al. 1985; Walker 2007a,b). As degeneration of the brainstem nuclei did not correlate with the Vonsattel HD grade of neostriatal atrophy, it is likely that damage to these subcortical nuclei does not develop simultaneously with, but independently from the well-known pathognomonic lesions in the neostriatum (Rüb et al. 2014a).

The *medial, lateral, and superior vestibular nuclei* via their descending vestibulospinal projections contribute substantially to the maintenance of postural mechanisms during stance and gait which normally prevent falls. Therefore, the involvement of these vestibular nuclei together with that of the cerebellum and the interconnected ventrolateral thalamic nucleus sufficiently explains why HD patients also may suffer from gait and truncal and postural instability and show a tendency to fall (Fig. 5.7) (Busse et al. 2009; Dom et al. 1976; Grimbergen et al. 2008; Rüb et al. 2004a, 2008b, 2009, 2013a, 2014a).

The *pontine nuclei* and the *inferior olive* are assigned to the precerebellar brainstem nuclei and are intimately connected with the cerebellum (ponto- or neocerebellum: pontine nuclei and inferior olive) and vermal and paravermal regions of the cerebellum (vestibulo- or archicerebellum, spino- or paleocerebellum: inferior olive), which has been recently demonstrated to be among the consistent targets of the neuropathological process of HD. Owing to their interconnectivities with the cerebellum, these brainstem nuclei together with a variety of other so-called precerebellar brainstem nuclei act as a gate for all afferent input from the neocortex, brainstem, and spinal cord destined for the cerebellum and are integrated into neuronal motor feedback loops, which ultimately control the activity of premotor and motor neurons in the lower brainstem and spinal cord (Fig. 5.7) (Braak et al. 2003b; Rüb et al. 2005, 2014a). Accordingly, loss of Purkinje cells in the cerebellar cortex and neurodegeneration of the four deep cerebellar nuclei of HD patients in combination with neuronal loss in the precerebellar ventrolateral thalamic nucleus, pontine nuclei, and inferior olive most likely are detrimental to the functions accomplished by the *precerebellar and cerebellar neural circuits* and thus contribute substantially to the cerebellar disease symptoms observed in HD patients (i.e., impairments of rapid alternating movements and fine motor skills, dysarthria, hypotonia, postural instability and ataxia, impaired gait balance, falls, broad-based gait and stance) (Busse et al. 2009; Dom et al. 1976; Grimbergen et al. 2008; Knikou 2008; Koller and Trimble 1985; Margolis and Ross 2003; Rüb et al. 2005, 2013a, 2014a; Scherzed et al. 2012; Walker 2007a, b).

Embedded into the trapezoid body, the *superior olive* with its three subnuclei (Fig. 7.2) is crucial for the performance of auditory functions (i.e., perception of the temporal patterns of sounds and analysis of the location, distance, and movement of sound sources) (Hoche et al. 2008; Moore and Linthicum 2004; Rüb et al. 2014a). Accordingly, the severe and consistent involvement of the superior olive in HD patients suggests that *auditory dysfunctions*, such as impairments of perception of the frequency, intensity, and temporal patterns of sounds, disturbed analysis of attributes of sound sources (i.e., location, distance, and movement), impaired discriminative signal/noise hearing, impaired speech discrimination, or pathologically altered *brainstem auditory-evoked potentials* (*BAEP*) may represent common HD disease signs. However, the findings of BAEP investigations of HD patients are still contradictory (Josiassen et al. 1984; Lin et al. 2011), and there is a striking discrepancy between the consistent involvement of the superior olive and the absence of clinical reports of impaired hearing in HD patients. Explanations that may account for the rarity of such clinical reports in HD include the following: (1) The clinical picture of HD is dominated by motor symptoms, psychiatric manifestations, and cognitive decline. Accordingly, more subtle auditory disease symptoms may often escape the attention of the affected patients and/or their caregivers. (2) Clinicians still are unaware of the involvement of auditory brainstem nuclei in HD; therefore, they do not expect auditory symptoms and are not inclined to perform detailed clinical or extensive neurophysiological examinations of auditory functions.

Degeneration of the dopaminergic compact part of the *substantia nigra* (Fig. 7.1) is a characteristic and unifying neuropathological feature of all neurological diseases that come along with parkinsonism (i.e., Parkinson's disease, PD; progressive supranuclear palsy, PSP; multiple system atrophy, MSA; corticobasal degeneration, CBD) (Den Dunnen 2013; Dickson 1999; Halliday 2004; Paulus and Jellinger 1991; Rüb et al. 2014a). Although a consistent loss of nerve cells occurs in the compact and reticulate parts of the substantia nigra of HD patients, the severity and extent of the neuronal loss in the compact part of the substantia nigra present in HD may not suffice to provoke clinically relevant parkinsonian features in adult HD patients. This assumption is in agreement with the clinical experience that parkinsonian extrapyramidal motor features (e.g., rigidity) only emerge in *HD patients with juvenile onset* (Lange and Aulich 1986; Lange et al. 1976; Oyanagi et al. 1989; Rüb et al. 2014a; Vonsattel et al. 1985; Walker 2007a, b).

This chapter was written by Udo Rüb.

Chapter 8
Intraneuronal Transport and Defense Mechanisms with Possible Pathogenetic Relevance in Huntington's Disease (HD)

8.1 Axonal Transport Mechanisms in Healthy Nerve Cells

Neurons are extremely polarized cells with a complex and unique morphology, as well as long processes (axon and dendrites) that extend far away from the nerve cell body. This architectonic feature of polarization with afferent (i.e., dendrites) and efferent processes (i.e., axons) requires efficient communication between the body of nerve cells and their periphery and makes nerve cells particularly dependent on functionally intact, sufficient, and timely axonal and/or dendritic intracellular transport processes over long distances. These *intra-axonal trafficking processes* guarantee bidirectional transport of subcellular compartments, including membrane organelles to or from the nerve cell processes, and enable the supply for specific action sites in the axon (e.g., nodes of Ranvier, axon terminal, synapses) and the disposal of vital cell organelles, membrane structures, synaptic vesicles, soluble neuronal proteins, or ion channels. Therefore, these targeted intraneuronal transport processes represent an essential prerequisite for the survival of the nerve cells of the human brain. The functional maintenance of nerve cells depends largely on the sufficient, locally correct, and precisely timed anterograde axonal transport of proteins and cell organelles from the cell body. In addition, the proper nerve cell function requires the retrograde trafficking of worn-out cell organelles back from the terminal regions to the cell body (Fig. 8.1) (De Vos et al. 2008; Gunawardena and Goldstein 2005; Li and Conforti 2013; Millecamps and Julien 2013). Long-range *microtubule-based transport of cargos* is the main mechanism (1) to deliver and target cellular components, organelles, and proteins from the cell body to their peripheral action site and/or (2) to remove them in the case of exhaustion. This long-range microtubule-based transport is accomplished by two major components: (1) "engines" or main microtubule-based *molecular motors* (i.e., kinesin, dynein) which need energy from ATP hydrolysis to move *cargos* along the microtubule

U. Rüb et al., *The Neuropathology of Huntington's Disease: Classical Findings, Recent Developments and Correlation to Functional Neuroanatomy*,
Advances in Anatomy, Embryology and Cell Biology, Vol. 217,
DOI 10.1007/978-3-319-19285-7_8

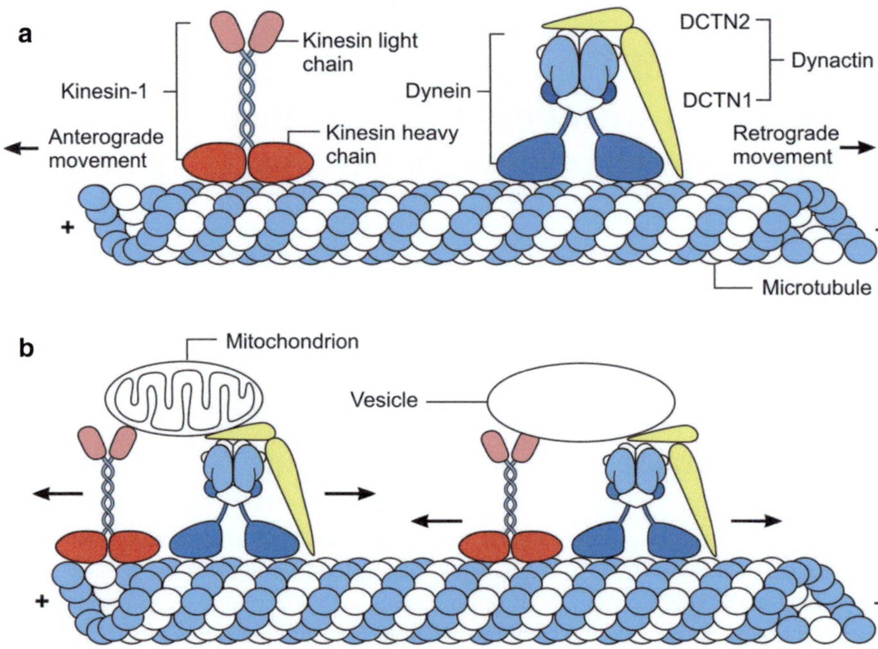

Fig. 8.1 The axonal transport mechanisms in healthy nerve cells. (**a**) The long-range *microtubule-based axonal transport of cargos* is the main mechanism to deliver and target cellular components, organelles, and proteins from the neuronal cell body to the action sites in the periphery and/or to remove them. This transport is accomplished by "engines" or main microtubule-based *molecular motors* (i.e., kinesin or dynein which generates force from ATP hydrolysis to move *cargos* along the microtubule cytoskeletal tracks) and *neuronal microtubules*, which represent the "rails" on which cargo-loaded molecular motors can run. The multimeric kinesin and dynein complexes have ATPase activity, can bind to the neuronal microtubule rails or tracks as well as to different cargos, and therefore represent the engines or molecular motors of the axonal transport that govern the microtubule-based axonal transport of most cargos and propel these cargos along neuronal microtubule cytoskeletal tracks. *Kinesin-1* is composed of two heavy chains (*red*) and two light chains (*light red*) and moves cargos in the anterograde direction along axons (i.e., toward the axon tip and the plus end of microtubules, +). The heavy chains of kinesin-1 are able to bind to microtubule rails and contain its ATPase activity, while its light chains bind the cargos. *Cytoplasmic dynein* is a multi-subunit protein complex that moves cargos in the retrograde direction (i.e., toward the minus end of the microtubules, –) and is built up by heavy chains (*blue*) and intermediate, light intermediate, and light chains (*light blue*), as well as by dynactin subunits (such as DCTN1 and DCTN2; *yellow*) implicated in cargo binding. While the heavy dynein chains harbor ATPase activity and bind to neuronal microtubules, all other protein chains of dynein are involved in cargo and dynactin binding. (**b**) Cargos (e.g., mitochondria and vesicles) are attached to the molecular motor kinesin-1 by its light chains (light red) and to the engine dynein by its intermediate, light intermediate, and light (*light blue*) chains, as well as by associated dynactin subunits (*yellow*) (Modified according to Millecamps and Julien (2013), (Figure 1, page 163); with kind permission from Nature Publishing Group). Abbreviations: *DCTN1* dynactin subunit 1, *DCTN2* dynactin subunit 2

cytoskeletal tracks and (2) *neuronal microtubules*, which represent the "rails" on which molecular motors loaded with cargos can run. Microtubules are very important in a number of cellular processes and are involved in maintaining the structure of the cell and, together with microfilaments and intermediate filaments as well as microtubules, form the cytoskeleton within the nerve cell's cytoplasm. Neuronal microtubules are long, hollow cylinders made up of polymerized *α- and β-tubulin dimers*, are among the core structures of the neuronal cytoskeleton, and are crucial for the maintenance of its structure as well as for organizing motility and transport of intracellular constituents. *Microtubule-associated proteins* (*MAP*) like the cytoskeletal protein tau bind to and stabilize neuronal microtubules, promote their correct polymerization and assembly, and maintain their structural integrity. Neuronal microtubules are highly dynamic structures capable to undergo rapid periods of growth and shrinkage to generate force and enable the motor proteins kinesin and dynein to transport neuronal organelles and other cellular components, as well as neuronal proteins (Fig. 8.1) (De Vos et al. 2008; Gunawardena and Goldstein 2005; Li and Conforti 2013; Millecamps and Julien 2013).

The *molecular motors kinesin and dynein* have ATPase activity, can bind to the neuronal microtubule rails as well as to different cargos, and are crucial role for the movement of cargos along cytoskeleton tracks. *Kinesin-1* is the best known member of the kinesin superfamily, consists of two kinesin heavy and two kinesin light chains, and harbors a catalytic motor domain. This motor domain binds to microtubule rails, contains the ATPase activity, and together with a short neck linker region confers processivity and directionality. The tail of kinesin-1 together with its light chains binds the cargo and thus regulates the motor activity of kinesin-1. *Cytoplasmic dynein*, likewise, is a multi-subunit protein complex built up by two heavy chains, as well as intermediate, light intermediate, and light chains. While the heavy dynein chains harbor ATPase activity and, via the motor domains, can bind neuronal microtubules, all other dynein protein chains are implicated in cargo and dynactin binding (Fig. 8.1).

Most of the known molecular motors are involved in cargo trafficking in only one direction. While molecular motors of the kinesin family mediate axonal transport toward the synapse (i.e., *anterograde axonal transport*), molecular motors of the dynein family mediate transport of most cargos toward the cell body (i.e., *retrograde axonal transport*). Based on the bulk speeds of saltatory cargo movement, the axonal transport mechanisms in both anterograde and retrograde directions are classically divided into fast and slow components. While cargos such as vesicles, mitochondria, and endosomes move by *fast axonal transport* at speeds of ~1 μm/s, neuronal cytoskeletal components move in *slow axonal transport* at speeds of ~1 mm/day. Fast as well as slow components of anterograde and retrograde axonal transport processes are mediated by the same molecular motors kinesin and cytoplasmic dynein. Their different speeds result from different pauses between kinesin- and/or dynein-based cargo movements in the anterograde and retrograde directions (Fig. 8.1) (De Vos et al. 2008; Gunawardena and Goldstein 2005; Li and Conforti 2013; Millecamps and Julien 2013).

Damage to and *disruption of axonal transport in human neurodegenerative diseases* can occur via four mechanisms and contribute to dysfunctions and/or demise of affected nerve cells: (1) impairments of the engines or molecular motors, (2) damage to the microtubule rails, (3) damage to the cargos, and (4) loss of energy supply caused by mitochondrial damage (Fig. 8.2) (De Vos et al. 2008; Gunawardena and

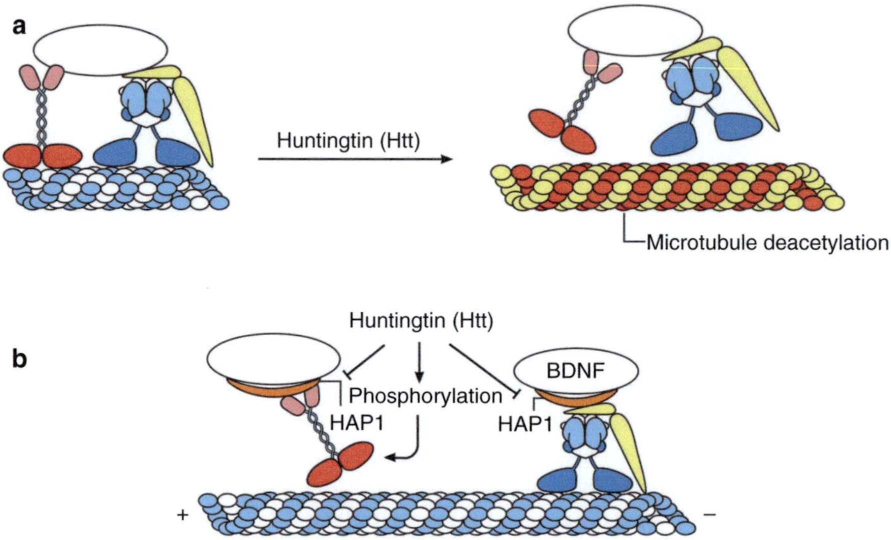

Fig. 8.2 Impairments of axonal transport mechanisms in Huntington's disease (HD). In human neurodegenerative diseases damage to and *disruption of axonal transport mechanisms* can occur via four mechanisms which ultimately contribute to dysfunctions and/or demise of affected nerve cells: (1) impairments of the engines or molecular motors, (2) damage to the microtubule rails, (3) damage to the cargos, (4) injury of the energy supply caused by mitochondrial damage. Although the exact pathophysiological mechanisms are not fully understood, the following functional restrictions of mutant huntingtin (*Htt*) are currently considered to contribute to impairments of axonal transport processes in Huntington's disease (HD): (1) reduced acetylation of the α-tubulin components of microtubules, which results in an impairment of kinesin-1 and dynein binding to and release of these molecular motors from the microtubule rails, (2) impaired binding and interaction of mutant Htt with the Htt-associated protein HAP1 that infers with axonal vesicular trafficking and organelle transport, (3) abnormal phosphorylation of kinesin-1 that reduces the binding of this molecular motor to and interaction with the microtubule tracks, and (4) abnormal anterograde and retrograde trafficking of mitochondria. (**a**) In HD, axonal transport defects may be caused by disruption of the binding of the motor proteins to microtubules or cargos. Mutant Htt promotes deacetylation of the α-tubulin components of microtubules and results in an impairment of kinesin-1 and dynein binding to and release of these molecular motors from the microtubule rails. (**b**) Microtubule track binding of kinesin-1 is also prevented by phosphorylation and interaction of mutant Htt with the Htt-associated protein HAP1 (*orange*). HAP1 is an adaptor protein that interacts with kinesin-1 light chains and the dynein-associated dynactin subunit 1 (DCTN1, *yellow*). Disruption of HAP1-mediated axonal transport also includes delivery of brain-derived neurotrophic factor (*BDNF*) (Modified according to Millecamps and Julien (2013), (Figure 2, page 16); with kind permission from Nature Publishing Group). Abbreviations: *BDNF* brain-derived neurotrophic factor, *HAP1* huntingtin-associated protein1, *Htt* huntingtin

Goldstein 2005; Li and Conforti 2013; Millecamps and Julien 2013). Mutations in the main microtubule-based motors kinesin-1 and cytoplasmic dynein which inhibit their activity as molecular carriers, as well as abnormal phosphorylation of kinesin-1, are among the known molecular causes that may lead to *impairments of the engines or molecular motors* of axonal transport mechanisms in human neurodegenerative diseases. In human neurodegenerative diseases *damage to microtubule rails* can result from inhibition, abnormal bundling or destabilization of microtubules, damage to the axonal trafficking of microtubules themselves, and/or impaired interactions of microtubules with molecular motors. Each of these damages can lead to the deregulation of the dynamic properties of microtubule rails and to the disruption of cargo transport via axonal transport mechanisms. Mechanisms that lead to axonal transport defects in neurodegenerative diseases via *damage to cargos and disruption of molecular motor-cargo interaction* include abnormal phosphorylation or misassembly of cargos (e.g., neurofilaments). Finally, axonal transport deficits caused by *mitochondrial damage* are associated with a reduced anterograde transport of mitochondria and vesicles which results in a diminution of mitochondria in axons and decrease of ATP supply to molecular motors and, in turn, leads to decreased anterograde and retrograde movement of other axoplasmic cargos. The reduced ATP production caused by mitochondrial dysfunction might be part of a vicious circle that ultimately leads to dying-back of axons (Fig. 8.2) (De Vos et al. 2008; Gunawardena and Goldstein 2005; Li and Conforti 2013; Millecamps and Julien 2013).

Mutiple evidences suggest that HD is among the polyglutamine diseases which go along with disruptions of axonal transport mechanisms and that damage to these transport processes is primarily associated with impaired functionality of *mutant huntingtin* (*Htt*). Although the exact pathophysiological mechanisms are not fully understood, the following functional restrictions of mutant Htt are currently considered to contribute to *impairments of axonal transport processes in HD*: (1) impaired binding and interaction of mutant Htt with the Htt-associated protein HAP1 that interferes with axonal vesicular trafficking and organelle transport, (2) reduced acetylation of the α-tubulin components of microtubules and consecutive impaired kinesin-1 binding and release of the molecular motor kinesin-1 from the microtubule rails, (3) abnormal phosphorylation of kinesin-1 which reduces its binding and interaction with the microtubule tracks, and (4) abnormal anterograde and retrograde trafficking of mitochondria (Fig. 8.2) (De Vos et al. 2008; Gunawardena and Goldstein 2005; Li and Conforti 2013; Millecamps and Julien 2013; Ortega et al. 2007).

8.2 The Neuronal Protein Quality Control Machinery: Molecular Chaperones and the Ubiquitin-Proteasome Pathway

Conformational changes and misfolding of specific disease proteins and their subsequent intraneuronal or extraneuronal aggregation are common molecular features shared by many neurodegenerative diseases including HD. Therefore, these human brain diseases are collectively referred to as *proteinopathies*, *protein-misfolding disorders*, or *protein-conformational disorders* (Atkin and Paulson 2014; Brundin et al. 2010; Goedert et al. 2010; Jucker and Walker 2011; Leak 2014; Ortega et al. 2007; Wooten et al. 2006). All types of intraneuronal and extraneuronal aggregations of conformationally changed and misfolded proteins which occur in human neurodegenerative diseases only develop within the framework of the ongoing disease process. They thus represent abnormal constituents of diseased brains. Since these aggregated, aberrant nerve cell constituents cannot be diluted by the *postmitotic nerve cells* through cell division, postmitotic neurons in particular rely on elaborated pathways of protein quality control and protein removal to prevent protein aggregation into pathological inclusions (Atkin and Paulson 2014; Ortega et al. 2007). Although the sequence of amino acids of a given protein contains all information necessary for its proper folding into a functional, three-dimensional structure, newly synthesized proteins may also undergo conformational change and misfolding, which together with a concomitantly decreased activity of the *neuronal protein quality control machinery* represents a risk factor for their intraneuronal accumulation and aggregation (Atkin and Paulson 2014; Brundin et al. 2010; Davies et al. 2007; Ortega et al. 2007; Paul 2008; Renner and Melki 2014; Wooten et al. 2006).

Protein misfolding can be prevented or even reversed by *molecular chaperones*. If the *molecular chaperone* activity fails, the proteins must be degraded before aggregation takes place. Even with the abundant activity of molecular chaperones, large amounts of newly synthesized proteins (i.e., approximately 30 %) are defective "off-pathway" products, misfolded due to mutations and conformational changes or inefficient assembly, and tend to aggregate in nerve cells (Borrell-Pagès et al. 2006; Imarisio et al. 2008; Leak 2014; Ortega et al. 2007; Wooten et al. 2006). Impairments of nascent proteins normally can be corrected by degradation of the misfolded proteins shortly after their synthesis by the *ubiquitin-proteasome pathway* (*UPS*) (Figs. 8.3 and 8.4).

The UPS works in the nucleus and cytoplasm of nerve cells, is highly conserved across species, and represents a major defense against short-lived and misfolded proteins in postmitotic nerve cells. Since postmitotic nerve cells are unable to divide and thus to reduce their burden of damaged proteins, the UPS plays a crucial role for protein turnover and degradation of damaged proteins (Atkin and Paulson 2014; Brundin et al. 2010; Imarisio et al. 2008; Leak 2014; Li and Li 2011; Ortega et al. 2007; Paul 2008; Wooten et al. 2006). Failure to remove damaged and polyubiquitinated proteins may lead to the accumulation of the pathologically changed proteins, as well as formation of aggregates and inclusion bodies in affected nerve cells, which are the pathognomonic hallmarks of many neurodegenerative diseases. Since aggregates formed in diseased nerve cells are more stable than the improperly folded proteins, effective unfolding of aberrantly proteins by means of molecular chaperones or degradation of misfolded substrates by the UPS into small peptides must gain the mastery over the neuronal aggregation process. The ultimate fate of misfolded proteins, however, depends on kinetic partitioning between these detrimental pathways and defense or protective mechanisms (Leak 2014; Li and Li 2011; Ortega et al. 2007; Paul 2008; Schulte and Littleton 2011; Wooten et al. 2006).

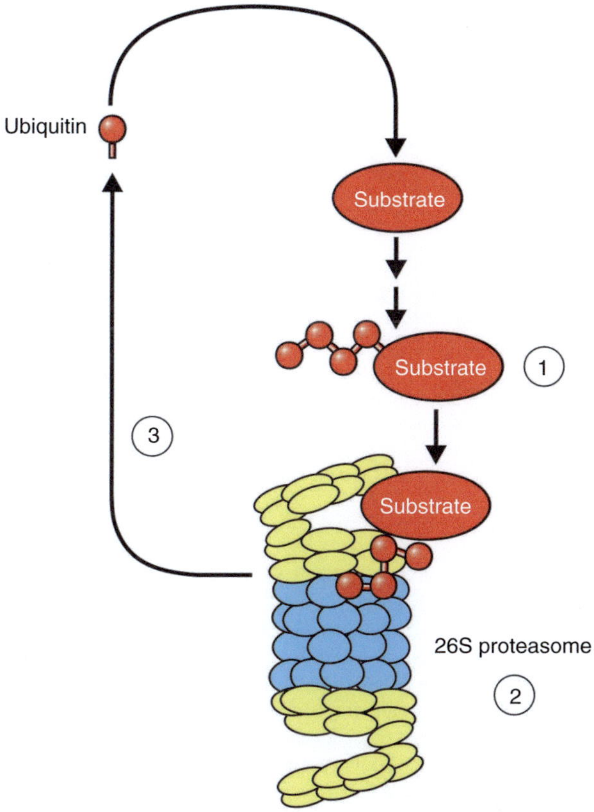

Fig. 8.3 Polyubiquitin tagging of conformationally changed and misfolded protein substrates for degradation by the 26S proteasome complex. *The ubiquitin-proteasome system* (*UPS*) serves as a cellular quality control system that tags misfolded proteins with polyubiquitin chains for degradation by the 26S proteasome: (*1*) covalent attachment of polyubiquitin chains to target the damaged protein substrates for degradation, (*2*) degradation of the polyubiquitin-tagged proteins by the proteolytic 26S *proteasome* complex, (*3*) de-ubiquitination of the protein substrate and release of the free and recyclable small ubiquitin monomers (Modified according to Wooten et al. (2006), (Figure 1, page 2); with kind permission from Hindawi Publishing Corporation)

To enhance the efficiency of correct de novo protein folding, to promote refolding of aberrantly folded proteins, and to prevent abnormal protein interactions and/or protein aggregation, nerve cells are equipped with a conserved class of proteins called *molecular chaperones* (Borrell-Pagès et al. 2006; Brundin et al. 2010; Imarisio et al. 2008; Labbadia and Morimoto 2013; Leak 2014; Li and Li 2011; Ortega et al. 2007; Paul 2008; Wooten et al. 2006). Molecular chaperones represent the first line of defense against conformationally changed and misfolded proteins prone to aggregate in nerve cells. They can transiently stabilize or refold such protein species, can mediate the folding or assembly of unfolded protein substrates, and are thus intimately associated with the intraneuronal mechanisms of protein

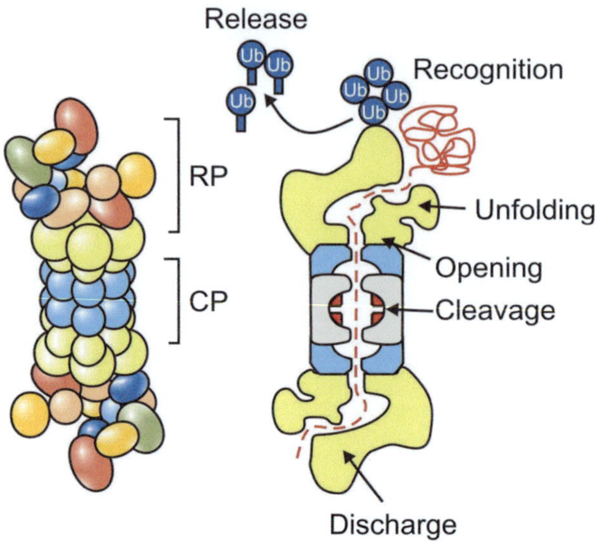

Fig. 8.4 The structure and function of the 20S core and 19S regulatory particle of the proteasome. The 26S proteasome is a multimeric protease and ATP-dependent complex which subserves protein degradation through ubiquitin-dependent and ubiquitin-independent mechanisms. It is composed of a proteolytically active *20S core particle* (*CP*, blue), as well as one or two *19S regulatory caps* (*RP*, yellow). The RP subserve recognition, unfolding, de-ubiquitination, and translocation of aberrant substrate proteins into the lumen of the CP. The CP by means of its proteolytic activity can finally degrade protein substrates that enter its catalytic chamber (Modified according to Wooten et al. (2006), (Figure 5, page 5); with kind permission from Hindawi Publishing Corporation). Abbreviations: *CP* 20S core protease, *RP* 19S regulatory particle, *Ub* ubiquitin

degradation (Borrell-Pagès et al. 2006; Brundin et al. 2010; Imarisio et al. 2008; Labbadia and Morimoto 2013; Leak 2014; Li and Li 2011; Ortega et al. 2007; Paul 2008; Wooten et al. 2006). Under pathological conditions (e.g., in human proteinopathies), when molecular chaperones cannot repair the consequences of massive protein misfolding, the subsequent depletion of molecular chaperones and the rising levels of aggregate-prone proteins can exceed the refolding capacity of the chaperone system. In this pathological scenario, two further mechanisms can be activated to eliminate and degrade misfolded proteins into small peptides: the chaperone-mediated targeting to the *ubiquitin-proteasome pathway* (*UPS*) or the *autophagy-lysosomal pathway* (*ALP*) (Figs. 8.3, 8.4, and 8.5) (Atkin and Paulson 2014; Brundin et al. 2010; Finkbeiner and Mitra 2008; Imarisio et al. 2008; Labbadia and Morimoto 2013; Leak 2014; Li and Li 2011; Ortega et al. 2007; Wooten et al. 2006).

While the UPS degrades most short-lived misfolded proteins into recycable peptides, the ALP represents a degradation process that refers to the lysosomal degradation of intracellular components, long-lived proteins, and organelles. Under physiological conditions of the collective activities of the molecular chaperones, the UPS and ALP suffice to handle mild fluctuations in the environmental milieu and to prevent the accumulation of conformationally changed and misfolded proteins. However, under certain

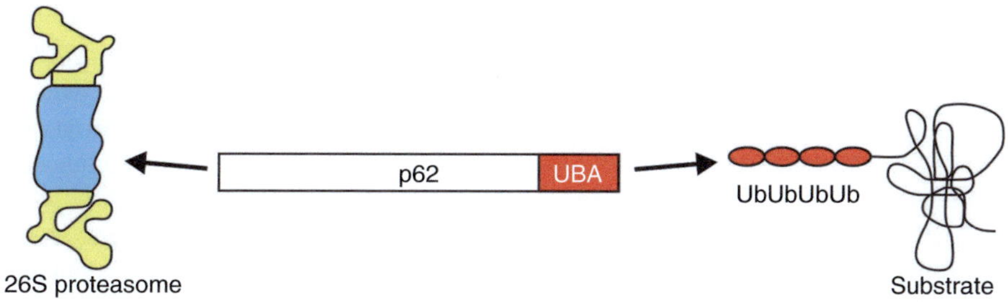

Fig. 8.5 The proteasomal shuttle and ubiquitin-interacting protein p62. The *proteasomal shuttle and ubiquitin-interacting protein p62* serves the presentation of polyubiquitin-tagged protein (*UbUbUbUb*) substrates to the 26S proteasome and interacts with polyubiquitinated protein substrates, as well as with the proteolytic proteasome. Its ubiquitin-association domain (*UBA*) binds covalently to and enables the low-affinity interaction of p62 with polyubiquitinated protein substrates (Modified according to Wooten et al. (2006), (Figure 6, page 5); with kind permission from Hindawi Publishing Corporation). Abbreviations: *Ub* ubiquitin, *UBA* ubiquitin-association domain

pathological conditions, the capacity of this protein quality control machinery is exceeded, and this failure results in the accumulation of dangerous levels of the misfolded protein (Atkin and Paulson 2014; Brundin et al. 2010; Finkbeiner and Mitra 2008; Imarisio et al. 2008; Labbadia and Morimoto 2013; Leak 2014; Li and Li 2011; Ortega et al. 2007; Paul 2008; Renner and Melki 2014; Wooten et al. 2006).

Heat shock proteins (*HSP*) are a group of stress proteins induced by heat shock; they are classified on the basis of their approximate molecular mass into six main families (Borrell-Pagès et al. 2006; Brundin et al. 2010; Imarisio et al. 2008; Labbadia and Morimoto 2013; Leak 2014; Li and Li 2011; Ortega et al. 2007; Paul 2008; Wooten et al. 2006). The activities of HSP are an integral part of the nerve cell's own repair system and neuronal protein quality control machinery and of the so-called "cellular stress response" or "heat shock response." The expression of HSP is upregulated when cells are exposed to different kinds of environmental stress conditions (e.g., infection, inflammation, exposure to toxins, starvation). The most prominent members of HSP are a class of functionally related proteins that function as intracellular molecular chaperones that simply "monitor" the cell's proteins and play an important role in protein-protein interactions, such as folding and assisting in the establishment of proper protein conformation and in the prevention of unwanted protein aggregation. They can help to stabilize partially unfolded proteins, are involved in intracellular protein trafficking, as well as coping with proteins denatured by heat and other stress forms (Borrell-Pagès et al. 2006; Brundin et al. 2010; Imarisio et al. 2008; Labbadia and Morimoto 2013; Leak 2014; Li and Li 2011; Ortega et al. 2007; Paul 2008; Wooten et al. 2006).

The *UPS* serves as a cellular quality control system that tags misfolded proteins with *ubiquitin chains* for degradation by the 26S proteasome. This process involves three crucial steps: covalent attachment of polyubiquitin chains to target the damaged proteins for degradation, degradation of the polyubiquitin-tagged proteins by the *26S proteasome* complex via proteolysis, and substrate de-ubiquitination and release of the free and recyclable small ubiquitin monomers (Atkin and Paulson 2014; Borrell-Pagès et al. 2006; Brundin et al. 2010; Davies et al. 2007; Finkbeiner and Mitra 2008; Imarisio et al. 2008; Leak 2014; Li and Li 2011; Ortega et al. 2007; Paul 2008; Wooten et al. 2006). For recognition and subsequent degradation of the protein substrate by the 26S proteasome complex, the formation of a chain of at least four conjugated molecules of the 8 kDa peptide ubiquitin is required (Figs. 8.3, 8.4, and 8.5).

The ATP-dependent complex of the 26S proteasome represents a multimeric protease complex and subserves protein degradation through both ubiquitin-dependent and ubiquitin-independent mechanisms. It is composed of a proteolytically active *20S core particle*, as well as one or two *19S regulatory caps*, which accomplish recognition, unfolding, de-ubiquitination, and translocation of substrate proteins into the lumen of the 20S core particle. The barrel-shaped 20S core particle consists of four stacked rings with proteolytic activity. Protein substrates that enter the catalytic chamber of the 20S proteasome can finally be degraded by its proteolytic activity (Figs. 8.3 and 8.4) (Atkin and Paulson 2014; Borrell-Pagès et al. 2006; Brundin et al. 2010; Davies et al. 2007; Finkbeiner and Mitra 2008; Li and Li 2011; Ortega et al. 2007; Paul 2008; Wooten et al. 2006). The physiological capacity of the UPS can be exceeded or overstretched either by overexpression of substrates or by a decrease in proteasome activity, and this might be a factor that initiates the formation of neuronal protein aggregation, which in turn may lead to a further decline of the UPS activity (Atkin and Paulson 2014; Borrell-Pagès et al. 2006; Brundin et al. 2010; Davies et al. 2007; Finkbeiner and Mitra 2008; Imarisio et al. 2008; Labbadia and Morimoto 2013; Leak 2014; Li and Li 2011; Ortega et al. 2007; Paul 2008; Schulte and Littleton 2011; Walker 2007a, b).

8.3 The Proteasomal Shuttle and Ubiquitin-Interacting Protein p62

The *proteasomal shuttle and ubiquitin-interacting protein p62* is an essential component of the UPS and, together with ubiquitin, plays a central role in the highly dynamic pathways that remove and degrade short-lived or aberrantly structured proteins, thus preventing their aggregation. p62 is involved in the presentation of polyubiquitinated protein substrates to the 26S proteasome and interacts with polyubiquitinated substrates and the degradative proteasome. Like other ubiquitin-interacting shuttle proteins, p62 has an ubiquitin-association domain (UBA) that can directly bind covalently to ubiquitin chains and enable its low-affinity interaction with these chains. The low-affinity interaction between p62 and ubiquitin guarantees the rapid assembly and disassembly between polyubiquitin chains and the shuttle protein p62. As with other proteasomal shuttle proteins, p62 in its N-terminus also harbors a ubiquitin-like domain (UbL) that binds to and directly interacts with the proteasome (Fig. 8.5). (Kuusisto et al. 2001, 2008; Ortega et al. 2007; Rüb et al. 2014a; Seibenhener et al. 2004; Seidel et al. 2009, 2010; Wooten et al. 2006).

This chapter was written by Udo Rüb and Horst-Werner Korf.

Chapter 9
The Disease Protein Huntingtin and Neuronal Protein Aggregations in Huntington's Disease (HD)

9.1 The Disease Protein Huntingtin

Identified in 1993 and located on chromosome 4p16.3, the human HD gene (also called *IT15*) contains 67 exons and spans more than 200 kb. It harbors meiotically unstable *CAG trinucleotide or polyglutamine repeats* (cytosine, adenine, guanine) in its exon 1, which encode the very large 350 kDa *huntingtin protein* (*Htt*). Htt consists of about 3100 amino acids and undergoes extensive posttranslational modifications (Fig. 9.1) (Andrew et al. 1993; Atkin and Paulson 2014; Borrell-Pagès et al. 2006; Brundin et al. 2010; Cattaneo et al. 2005; Davies et al. 2007; Duyao et al. 1993; Finkbeiner and Mitra 2008; Imarisio et al. 2008; Labbadia and Morimoto 2013; Li and Conforti 2013; Li and Li 2011; Margolis and Ross 2003; Myers et al. 1991; Ortega et al. 2007; Paul 2008; Renner and Melki 2014; Schulte and Littleton 2011; The Huntington's disease Collaborative Research Group 1993; Vonsattel 2008; Walker 2007a, b).

In symptomatic HD patients and asymptomatic gene carriers, the CAG trinucleotide and the polyglutamine repeats are pathologically expanded and give rise to an *elongated polyglutamine tract* at the N-terminus of mutant Htt that confers the tendency to Htt to form intraneuronal inclusions (Fig. 9.1) (Andrew et al. 1993; Atkin and Paulson 2014; Borrell-Pagès et al. 2006; Cattaneo et al. 2005; Duyao et al. 1993; Finkbeiner and Mitra 2008; Gunawardena and Goldstein 2005; Imarisio et al. 2008; Labbadia and Morimoto 2013; Li and Conforti 2013; Li and Li 2011; Margolis and Ross 2003; Myers et al. 1991; Ortega et al. 2007; Paul 2008; Renner and Melki 2014; Schulte and Littleton 2011; The Huntington's disease Collaborative Research Group 1993; Vonsattel 2008; Walker 2007a, b; Wooten et al. 2006). The normal physiological CAG repeat sequences comprise 6–35 CAG triplets, whereby sequences of 28 and more already behave unstable during meiosis and may be prone to potentially symptomatic mutations. CAG repeat sequences longer than 35 are considered pathologically expanded, 36–40 CAG repeats lead to an incomplete penetrance, and expansions of 41 or more CAG repeats lead to the fully developed

U. Rüb et al., *The Neuropathology of Huntington's Disease: Classical Findings, Recent Developments and Correlation to Functional Neuroanatomy*,
Advances in Anatomy, Embryology and Cell Biology, Vol. 217,
DOI 10.1007/978-3-319-19285-7_9

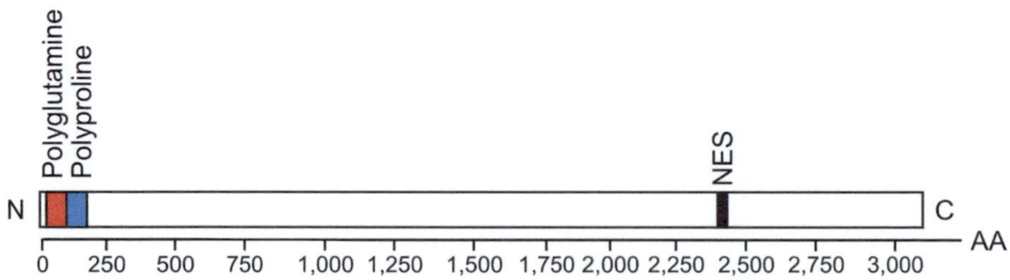

Fig. 9.1 The gene of the polyglutamine disorder Huntington's disease (HD). The human *Huntington* (*HD*) *gene* (also called *IT15* gene) contains 67 exons and spans more than 200 kb. In its N-terminus (*N*) it harbors meiotically unstable CAG trinucleotide repeats (*red*) which encode for the large polyglutamine protein huntingtin (Htt). In affected Huntington's disease (HD) patients and asymptomatic gene carriers, the CAG trinucleotide or polyglutamine repeats at the N-terminus (*N*) are pathologically expanded and give rise to an elongated polyglutamine tract in mutant Htt that confers the tendency of Htt to form insoluble intraneuronal inclusions. Depending on the length of the polyglutamine tract, Htt consists of about 3100 amino acids. Although the elongated polyglutamine tract at the N-terminus (*N*) of mutant Htt in higher vertebrates is flanked by a polyproline domain (*blue*) which is thought to help to maintain solubility of this protein, elongations of the polyglutamine tract in mutant Htt result in conformational changes, misfolding, and reduced solubility of Htt. The C-terminus (*C*) of Htt has an active nuclear export signal (*NES*) (*black*) which serves shuttling transcription factors between the nucleus and the cytoplasm of nerve cells (Modified according to Cattaneo et al. (2005) (Figure 1, page 922); with kind permission from Nature Publishing Group). Abbreviations: *AA* amino acid, *C* C-terminus, *Htt* huntingtin, *N* N-terminus, *NES* nuclear export signal

clinical picture of HD (Andrew et al. 1993; Atkin and Paulson 2014; Borrell-Pagès et al. 2006; Brundin et al. 2010; Cattaneo et al. 2005; Duyao et al. 1993; Finkbeiner and Mitra 2008; Imarisio et al. 2008; Labbadia and Morimoto 2013; Margolis and Ross 2003; Myers et al. 1991; Ortega et al. 2007; Schulte and Littleton 2011; The Huntington's disease Collaborative Research Group 1993; Walker 2007a, b).

The translated *wild-type polyglutamine protein Htt* with its polymorphic stretch of between 6 and 35 glutamines in its N-terminal domain represents a large 350-kDa protein, which may traffic between the nucleus and the cytoplasm of nerve cells. Htt shows a very high degree of conservation among vertebrates, but no homologies with other proteins. Htt is expressed ubiquitously throughout the human body with its highest levels in the brain and testes, is found in all neurons and all glial cells of the brain, and has its main brain expression sites in the cerebral neocortex, cerebellar cortex, striatum, and hippocampus (Borrell-Pagès et al. 2006; Cattaneo et al. 2005; Finkbeiner and Mitra 2008; Imarisio et al. 2008; Labbadia and Morimoto 2013; Li and Li 2011; Millecamps and Julien 2013; Ortega et al. 2007; Schulte and Littleton 2011; Vonsattel 2008; Walker 2007a, b).

The *subcellular neuronal localization* of the polyglutamine protein Htt is complex and dynamic. Htt may change its conformation depending on its compartmental localization in nerve cells, is mainly found in the cytoplasm of nerve cells, and co-localizes with many organelles including the nucleus, endoplasmic reticulum, Golgi complex, mitochondria, and endosomes. Htt is also observed in axonal processes, at synapses, as well as in association with microtubules, synaptic and autophagic vesicles, caveolae, and synaptosomes (Atkin and Paulson 2014; Borrell-Pagès et al.

2006; Cattaneo et al. 2005; Davies et al. 2007; Imarisio et al. 2008; Labbadia and Morimoto 2013; Li and Conforti 2013; Li and Li 2011; Millecamps and Julien 2013; Pandey et al. 2010; Schulte and Littleton 2011).

Htt interacts with many proteins that regulate endocytosis. Owing to its association with the endocytosis proteins clathrin and dynamin, as well as with endocytic organelle trafficking proteins, Htt is also involved in the short-range transport along the actin cytoskeleton as a component of the *endocytotic pathway* and as such is thought to allow Htt-associated endocytic vesicles to move along actin filaments.

The C-terminus of Htt has an active *nuclear export signal* (*NES*), while its N-terminus domain forms an amphipathic alpha-helical membrane-binding domain that reversibly mediates the association of Htt with the endoplasmic reticulum, endosomes, and autophagic vesicles (Fig. 9.1). Htt has *antiapoptotic activity* and a large number of *protein-protein interaction domains*. Owing to these domains it can interact with over 200 other cellular proteins including those involved in gene expression, intracellular signaling, intracellular transport of signaling molecules and trophic factors, as well as metabolism. Htt also appears to be involved in various functions in the neuronal cytoplasm and nucleus, binds to and interacts with various transcription factors, and may regulate transcription by shuttling transcription factors between the nucleus and the cytoplasm using its NES and by interacting with spliceosome-related proteins (Atkin and Paulson 2014; Borrell-Pagès et al. 2006; Davies et al. 2007; Imarisio et al. 2008; Millecamps and Julien 2013; Schulte and Littleton 2011).

The Htt protein also acts as a scaffold that links transport cargos with the molecular motor proteins of the *axonal transport* and additionally regulates factors that coordinate trafficking and transport of cellular material along and between actin and microtubule cytoskeletons in a bidirectional manner over both short and long distances. One of the main proteins interacting with Htt is the huntingtin-associated protein HAP1, which helps to mediate the interaction between Htt and the molecular motor proteins of anterograde and retrograde axonal transport mechanisms (see Sect. 8.1) (Fig. 8.2).

Moreover, evidence is also available that Htt is involved in the *sorting mechanisms of many proteins* at the Golgi region (Atkin and Paulson 2014; Borrell-Pagès et al. 2006; Cattaneo et al. 2005; Davies et al. 2007; De Vos et al. 2008; Gunawardena and Goldstein 2005; Imarisio et al. 2008; Labbadia and Morimoto 2013; Li and Conforti 2013; Li and Li 2011; Millecamps and Julien 2013; Ortega et al. 2007; Schulte and Littleton 2011). Finally, Htt is associated with the *antero- and retrograde movement of mitochondria* along axons of nerve cells (Figs. 8.1 and 8.2).

Along with that of other cargos, the anterograde and retrograde trafficking of mitochondria in axons may be impeded and physically blocked by intraneuronal protein aggregates in HD. This early disruption of mitochondrial trafficking in nerve cells leads to a reduced deposition of mitochondria at sites with a high local energy demand such as synapses; reduces the number of mitochondria at essential subcellular locations and their sufficient, timely, and specific local energy supply; and ultimately results in an impaired synthesis of ATP. Since both axonal trafficking and mitochondrial dysfunction may occur in HD, the links between these two processes during disease pathogenesis may be important for considerations of therapeutic intervention (see Sect. 8.1) (Fig. 8.2) (Borrell-Pagès et al. 2006; Davies et al. 2007;

De Vos et al. 2008; Gil and Rego 2008; Gunawardena and Goldstein 2005; Imarisio et al. 2008; Labbadia and Morimoto 2013; Li and Conforti 2013; Li and Li 2011; Millecamps and Julien 2013; Oliveira 2010; Ortega et al. 2007; Pandey et al. 2010; Rüb et al. 2013a, 2014a; Schulte and Littleton 2011; Walker 2007a, b).

Although the *elongated polyglutamine tract* of mutant Htt in higher vertebrates is flanked by a polyproline domain which is thought to help to maintain solubility of this protein (Fig. 9.1), elongations of the polyglutamine tract in mutant Htt result in conformational changes, misfolding, and a reduction of the solubility of this protein. All these processes most likely overload the defense mechanisms of affected nerve cells (i.e., molecular chaperones, UPS) necessary for cellular homeostasis of protein recycling and energetics and eventually lead to the formation of insoluble *ubiquitinated neuronal protein aggregations* (Atkin and Paulson 2014; Borrell-Pagès et al. 2006; Brundin et al. 2010; Cattaneo et al. 2005; Davies et al. 2007; Finkbeiner and Mitra 2008; Imarisio et al. 2008; Jucker and Walker 2011; Labbadia and Morimoto 2013; Ortega et al. 2007; Schulte and Littleton 2011; Walker 2007a).

To date, the molecular basis of the formation of the abnormally ubiquitinated neuronal aggregations is only fragmentarily understood. However, it is well known that these aggregations that evolve during the course of the polyglutamine disease HD contain, along with the small degradation-related 8 kDa ubiquitin protein, also highly ordered amyloid fibers with high β-sheet structure content (Fig. 9.1) and numerous other proteins, including factors important for transcription and protein quality control (e.g., proteasome subunits and molecular chaperones) (Atkin and Paulson 2014; Becher et al. 1998; Borrell-Pagès et al. 2006; Finkbeiner and Mitra 2008; Gunawardena and Goldstein 2005; Gutekunst et al. 1999; Imarisio et al. 2008; Labbadia and Morimoto 2013; Leak 2014; Li and Li 2011; Maat-Schieman et al. 1999; Ortega et al. 2007; Paul 2008; Schulte and Littleton 2011; Sieradzan et al. 1999; Walker 2007a, b; Wooten et al. 2006). Owing to the striking tendency of mutant Htt to form insoluble neuronal inclusions or aggregations, the pathological process of HD, like that of other known human proteinopathies or polyglutamine diseases, is associated with the occurrence of *insoluble neuronal aggregations* which represent the most striking immunocytochemical hallmarks of the HD-associated brain pathology (Atkin and Paulson 2014; Borrell-Pagès et al. 2006; Finkbeiner and Mitra 2008; Imarisio et al. 2008; Labbadia and Morimoto 2013; Li and Conforti 2013; Li and Li 2011; Maat-Schieman et al. 2007; Schulte and Littleton 2011; Walker 2007a, b; Wooten et al. 2006).

9.2 Types, Composition, and Pathogenetic Relevance of Neuronal Protein Aggregations in Huntington's Disease (HD)

In higher vertebrates the elongated polyglutamine tract of mutant Htt is flanked by a polyproline domain which is thought to help to maintain solubility of this protein (Fig. 9.1). The pathological *elongations of the polyglutamine tract in mutant Htt* result in conformational changes, misfolding, and a reduction of the solubility of the

Htt protein which most likely overload the defense mechanisms of the affected nerve cells (i.e., molecular chaperones, ubiquitin-proteasome pathway) necessary for cellular homeostasis of protein recycling and energetics and eventually lead to the formation of *insoluble ubiquitinated neuronal protein aggregations* (Atkin and Paulson 2014; Becher et al. 1998; Borrell-Pagès et al. 2006; Finkbeiner and Mitra 2008; Gunawardena and Goldstein 2005; Imarisio et al. 2008; Labbadia and Morimoto 2013; Leak 2014; Li and Li 2011; Maat-Schieman et al. 1999; Ortega et al. 2007; Schulte and Littleton 2011; Sieradzan et al. 1999; Walker 2007a, b; Wooten et al. 2006). Although, to date, the molecular basis of their formation is only fragmentarily understood, it is well known that the abnormal neuronal aggregations that evolve during the course of HD along with the small 76 amino acid protein *ubiquitin* also contain highly ordered amyloid fibrils with high β-sheet content and numerous other proteins, including factors important for transcription and components of the neuronal protein quality machinery (e.g., proteasome subunits and molecular chaperones) (Atkin and Paulson 2014; Becher et al. 1998; Borrell-Pagès et al. 2006; Davies et al. 2007; DiFiglia et al. 1997; Finkbeiner and Mitra 2008; Gunawardena and Goldstein 2005; Gutekunst et al. 1999; Imarisio et al. 2008; Labbadia and Morimoto 2013; Li and Li 2011; Maat-Schieman et al. 1999; Paul 2008; Schulte and Littleton 2011; Sieradzan et al. 1999; Walker 2007a, b; Wooten et al. 2006). Owing to the striking tendency of the mutated form of the polyglutamine protein Htt to form insoluble neuronal inclusions or aggregations, the pathological process of HD, like that of other known human proteinopathies or polyglutamine diseases, is associated with the occurrence of insoluble neuronal aggregations that represent the most prominent concomitant immunocytochemical hallmarks of the HD-associated brain pathology (Atkin and Paulson 2014; Becher et al. 1998; Borrell-Pagès et al. 2006; DiFiglia et al. 1997; Finkbeiner and Mitra 2008; Gil and Rego 2008; Gourfinkel-An et al. 1998; Gunawardena and Goldstein 2005; Gutekunst et al. 1999; Imarisio et al. 2008; Kuemmerle et al. 1999; Labbadia and Morimoto 2013; Li and Conforti 2013; Li and Li 2011; Maat-Schieman et al. 1999, 2007; Ortega et al. 2007; Paul 2008; Sapp et al. 1999; Schulte and Littleton 2011; Sieradzan et al. 1999; Valera et al. 2005; Walker 2007a, b).

Aggregations of Htt in the form of neuronal intranuclear inclusions (NI) were among the first types of neuronal aggregation pathologies described in HD, are well known as accompanying neuropathological feature of HD since more than two decades, and until recently were most frequently and nearly exclusively identified and described in the degenerated neostriatum and cerebral neo- and allocortex (Fig. 9.2) (Atkin and Paulson 2014; Becher et al. 1998; Borrell-Pagès et al. 2006; DiFiglia et al. 1997; Finkbeiner and Mitra 2008; Gil and Rego 2008; Gourfinkel-An et al. 1998; Gunawardena and Goldstein 2005; Gutekunst et al. 1999; Imarisio et al. 2008; Kuemmerle et al. 1999; Labbadia and Morimoto 2013; Li and Conforti 2013; Li and Li 2011; Maat-Schieman et al. 1999, 2007; Ortega et al. 2007; Rüb et al. 2014a; Sapp et al. 1999; Schulte and Littleton 2011; Sieradzan et al. 1999; Valera et al. 2005; Walker 2007b). Further biochemical studies showed that these subcortical and cortical NI along with the mutated disease protein Htt may also contain components of the cellular protein quality control machinery responsible for the handling, removing, and degrading of conformationally changed, misfolded, and polyubiquitin-tagged proteins (i.e., ubiquitin, proteasome subunits, molecular chaperones). Some previous neuropathological studies have demonstrated additional *neuropil* (i.e., *neuritic or axonal*) *aggregates* or swellings in select gray and white matter brain components of HD patients (i.e., neostriatum, cerebral neocortex, internal capsule, cerebral peduncle) (Atkin and Paulson 2014; Becher et al. 1998; Borrell-Pagès et al. 2006; Finkbeiner and Mitra 2008; Gil and Rego 2008; Gunawardena and Goldstein 2005; Gutekunst et al. 1999; Imarisio et al. 2008; Kuemmerle et al. 1999; Labbadia and Morimoto 2013; Li and Conforti 2013; Li and Li 2011; Maat-Schieman et al. 2007; Ortega et al. 2007; Paul 2008; Sapp et al. 1999; Schulte and Littleton 2011; Sieradzan et al. 1999; Valera et al. 2005; Walker 2007b; Wooten et al. 2006).

Recent *systematic immunocytochemical studies of the brainstem and cerebellum* of HD patients confirmed (1) that NI by no means represent the sole type of neuronal protein aggregations that develop during HD, (2) that NI are not confined to the classical brain regions known to undergo neurodegeneration in HD (i.e., neostriatum, cerebral cortex), and (3) that the neuronal protein aggregation pathology is more widespread in the brains of HD patients than hitherto thought (Figs. 9.2, 9.3, 9.4, 9.5, and 9.6) (Rüb et al. 2013a, 2014a). According to these studies (1) immunoreactive *compact and granular neuronal protein aggregations* may also occur in the cytoplasm of cerebellar nerve cells, (2) brainstem NI are present in both degenerated and spared nuclei of HD patients, and (3) the occurrence of *intra-axonal inclusions* in fiber tracts is a common and widespread feature of the cerebellar and brainstem pathology of HD patients (Figs. 9.2, 9.3, 9.4, 9.5, and 9.6) (Rüb et al. 2013a, 2014a).

Investigations of the *cerebellum* of HD patients disclosed fine and granular dispersed cytoplasmic protein aggregates in nearly all of the surviving *cerebellar Purkinje cells*, which were immunoreactive for the anti-polyglutamine antibody 1C2 that detects neuronal proteins with an expanded polyglutamine sequence of more than 37 glutamines (Rüb et al. 2014a; Seidel et al. 2010; Trottier et al. 1995). In some of the remaining Purkinje cells, additional dot-like NI immunoreactive for the 1C2 antibody were present (Fig. 9.6). Furthermore, immunocytochemistry for the calcium-binding protein calbindin that represents an acknowledged, selective, and highly reliable immunocytochemical marker for the large GABAergic cerebellar Purkinje cells revealed the presence of axonal inclusion bodies (so-called *torpedos*) in remaining Purkinje cells (Fig. 5.2) (Bastianelli 2003; Rüb et al. 2013a; Whitney et al. 2008). 1C2 immunocytochemistry, in addition, revealed neuronal aggregations with a presumed axonal localization in the cerebellar granular cell layer and underlying cerebellar white matter of HD patients (Rüb et al. 2013a). p62 immunoreactive neuronal inclusion bodies were also present in the surviving nerve cells of all four *deep cerebellar nuclei* of the HD patients studied. The overwhelming majority of these surviving nerve cells exhibited at least one of the three aggregate types (i.e., compact cytoplasmic or granular cytoplasmic aggregates, NI) (Fig. 9.6). In the remaining nerve cells of the fastigial and globose nuclei of the HD patients, compact cytoplasmic, granular cytoplasmic aggregates, and NI were equally

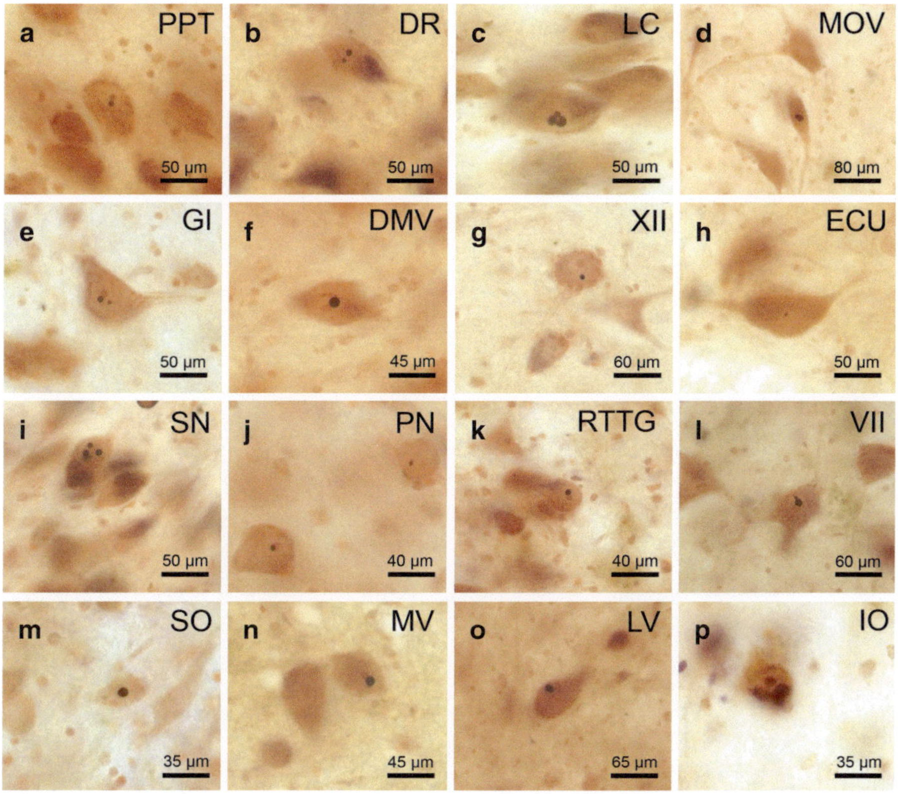

Fig. 9.2 Neuronal intranuclear inclusions in brainstem nuclei of Huntington's disease (HD) patients immunoreactive for the ubiquitin-interacting and proteasomal shuttle protein p62. *p62* immunoreactive neuronal intranuclear inclusions (NI) in *well-preserved brainstem nuclei* in Huntington's disease (HD): (**a**) pedunculopontine nucleus (*PPT*), (**b**) dorsal raphe nucleus (*DR*), (**c**) locus coeruleus (*LC*), (**d**) motor trigeminal nucleus (*MOV*), (**e**) gigantocellular reticular nucleus (*GI*), (**f**) dorsal motor vagal nucleus (*DMV*), (**g**) hypoglossal nucleus (*XII*), (**h**) external cuneate nucleus (*ECU*). p62 immunoreactive NI in *degenerated brainstem nuclei* in HD: (**i**) substantia nigra (*SN*), (**j**) pontine nuclei (*PN*), (**k**) reticulotegmental nucleus of the pons (*RTTG*), (**l**) facial nucleus (*VII*), (**m**) superior olive (*SO*), (**n**) medial vestibular nucleus (*MV*), (**o**) lateral vestibular nucleus (*LV*), (**p**) inferior olive (*IO*) (**a**, **b**, **f**, **l**, **n**, **o** HD patient with Vonsattel grade 4 of neostriatal atrophy; 4; **c**, **g**, **h**, **m** HD patient with Vonsattel grade of neostriatal atrophy; **d**, **i** HD patient with Vonsattel grade 4 of neostriatal atrophy; **e**, **j**, **k**, **p** HD patient with Vonsattel grade 3 of neostriatal atrophy) (**a–p**: p62 immunohistochemistry, counterstaining with aldehyde-fuchsin Darrow red, 100 µm polyethylene glycol sections) (Reprinted from Rüb et al. (2014a), (Figure 4, page 254); with kind permission from John Wiley and Sons)

frequent (Fig. 9.6). In contrast, in the nerve cells of the emboliform nucleus, compact cytoplasmic aggregates clearly dominated, while granular cytoplasmic aggregates and NI were only rarely encountered. In the dentate nucleus, granular cytoplasmic aggregates and compact cytoplasmic aggregates likewise were most prevalent, whereas NI were only rarely encountered (Fig. 9.6). Furthermore, anti-Htt

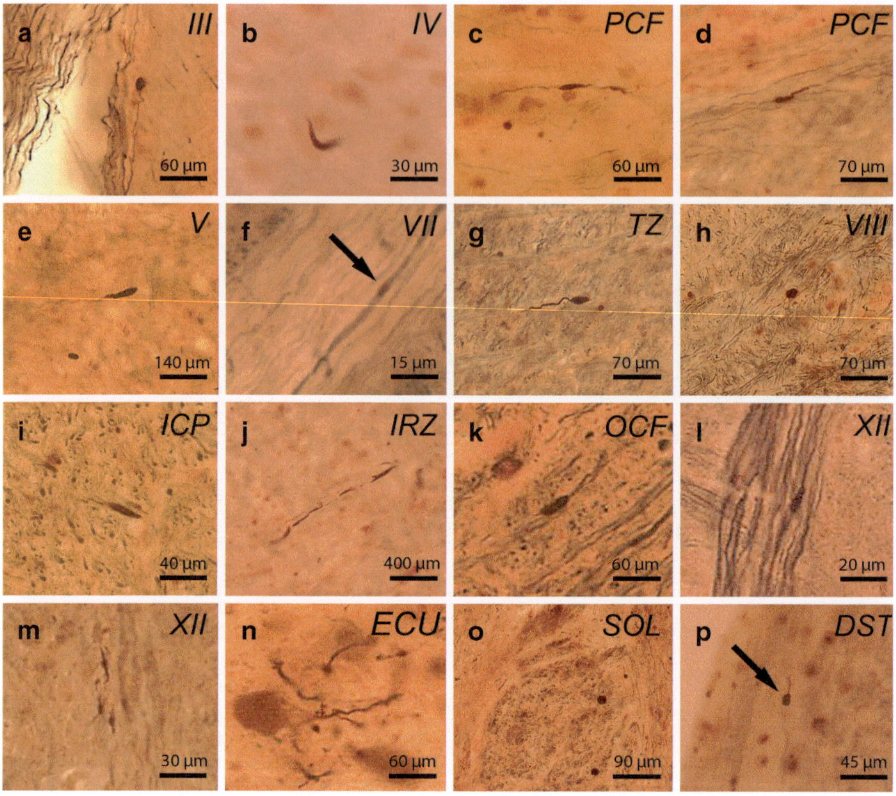

Fig. 9.3 Axonal inclusions in brainstem fiber tracts in Huntington's disease (HD) immunoreactive for the ubiquitin-interacting and proteasomal shuttle protein p62. *p62* immunoreactive *axonal inclusions* in Huntington's disease (HD): (**a**) oculomotor nerve (*III*), (**b**) trochlear nerve (*IV*), (**c**) pontocerebellar fibers (*PCF*), (**d**) *PCF*, (**e**) trigeminal nerve (*V*), (**f**) facial nerve (*VII*), (**G**) trapezoid body (*TZ*), (**h**) vestibulocochlear nerve (*VIII*), (**i**) inferior cerebellar peduncle (*ICP*), (**j**) intermediate reticular zone (IRZ), (**k**) olivocerebellar fibers (*OCF*), (**l**) hypoglossal nerve (*XII*), (**m**) hypoglossal nerve (*XII*), (**n**) cuneate fascicle (*ECU*), (**o**) solitary tract (*SOL*), (**p**) dorsal spinocerebellar tract (*DST*). (**a–p** p62/AT270 double immunohistochemistry, 100 µm PEG sections) (**a**, **e**, **k**, **n** HD patient with Vonsattel grade 4 of neostriatal atrophy; **b**, **l**, **p** HD patient with Vonsattel grade 2 of neostriatal atrophy; **c**, **d**, **f**, **g**, **h** HD patient with Vonsattel grade 3 of neostriatal atrophy; **i** HD patient with Vonsattel grade 3 of neostriatal atrophy; **j**, **o** HD patient with Vonsattel grade 3 of neostriatal atrophy: 3; **m** HD patient with Vonsattel grade 4 of neostriatal atrophy) (Reprinted from Rüb et al. (2014a), (Figure 5, page 255); with kind permission from John Wiley and Sons)

immunostaining disclosed a few compact neuropil aggregates in the molecular cell layer of the cerebellar cortex, numerous compact neuropil aggregates in the cerebellar granule cell layer and deep white matter, as well as a few NI in remaining Purkinje cells, and in the deep globose and dentate nuclei (Rüb et al. 2013a).

Recent studies also detected a widespread *p62 immunoreactive neuronal aggregation* pathology in the brainstem of HD patients (Figs. 9.2, 9.3, 9.4, and 9.5). p62

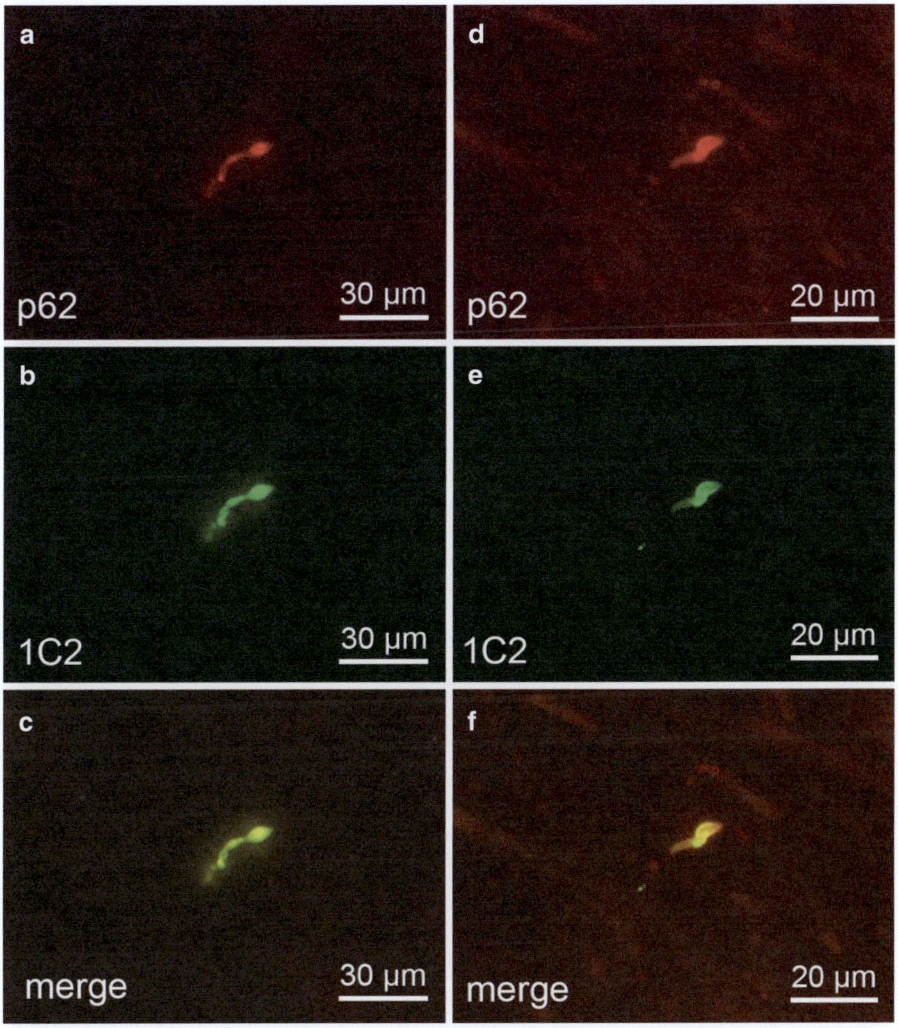

Fig. 9.4 Double immunofluorescence of axonal inclusions in brainstem fiber tracts in Huntington's disease (HD). (**a–c**) Double immunofluorescence of axonal inclusions in the nigrostriatal pathway and (**d–f**) in the pontocerebellar fibers of a representative Huntington's disease (HD) patient with Vonsattel grade 4 of neostriatal atrophy. (**a**, **d**) Immunolabeling of intraneuronal aggregates with an antibody against the ubiquitin-interacting and *proteasomal shuttle protein p62* (*red*) and (**b**, **e**) with the *anti-polyglutamine marker 1C2* (*green*). The positive immunoreactions for the anti-polyglutamine marker 1C2 point to the recruitment of the pathologically altered form of the disease protein huntingtin (Htt) with an elongated polyglutamine stretch into the p62 immunoreactive axonal inclusions. (**a**, **d** p62 immunostaining, Cy3; **b**, **e** anti-polyglutamine 1C2 immunostaining, Alexa 488, 100 µm polyethylene glycol sections) (Reprinted from Rüb et al. (2014a), (Figure 6, page 256); with kind permission from John Wiley and Sons)

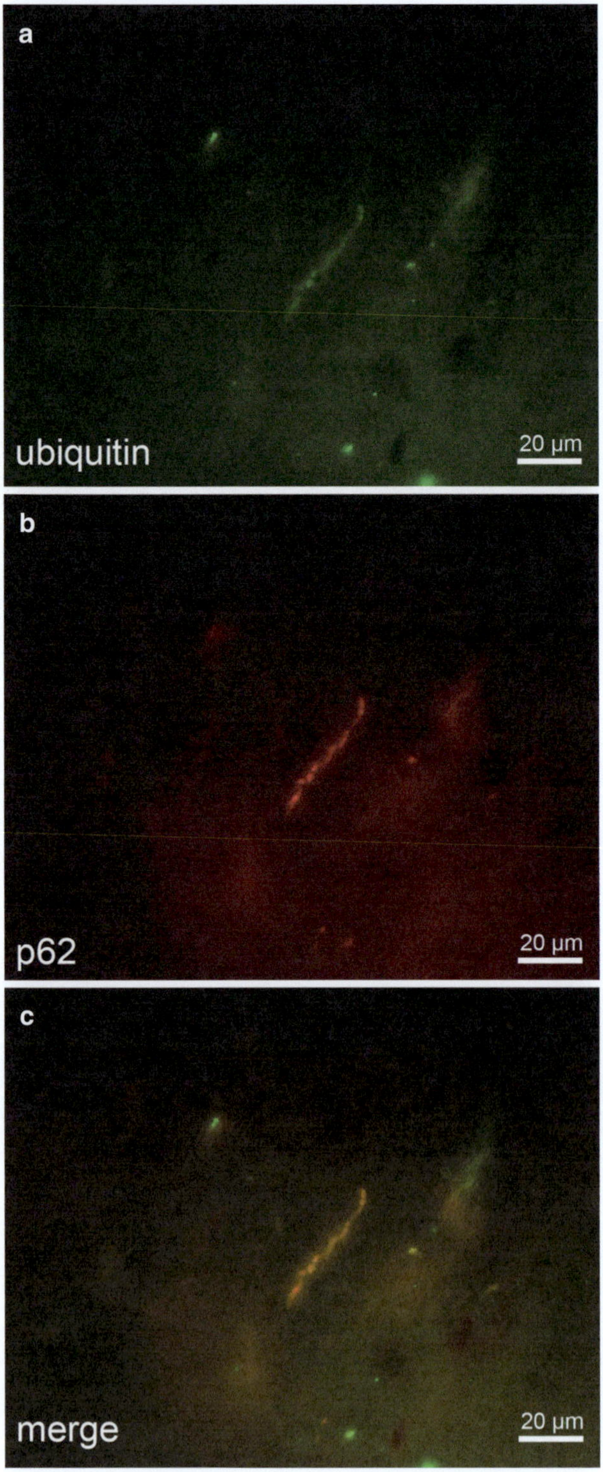

Fig. 9.5 Double immunofluorescence of axonal inclusions in the nigrostriatal pathway in Huntington's disease (HD). Double immunofluorescence of axonal inclusions in the nigrostriatal pathway of a representative Huntington's disease (HD) patient with Vonsattel grade 2 of neostriatal atrophy. (**a**) Immunolabeling of axonal inclusions with an antibody against the *ubiquitin* protein (*green*) and (**b**) the *proteasomal shuttle protein p62* (*red*). The presence of ubiquitin and p62 in axonal aggregates suggests an involvement of the ubiquitin-proteasome pathway (UPS) in the HD pathogenesis and points to defense mechanisms of affected nerve cells that attempt to reduce levels of mutant huntingtin (Htt), but may also reflect a decreased or compromised activity of the UPS or its overload in HD. (**a** p62 immunostaining, Alexa 488; **b** anti-ubiquitin immunostaining, Cy3; 100 μm polyethylene glycol sections) (Reprinted from Rüb et al. (2014a), (Figure 5, page 257); with kind permission from John Wiley and Sons)

immunoreactive *NI* were consistently present in all brainstem nuclei of HD patients, irrespective of whether they were a target of the destructive process or were well preserved. In some of the remaining brainstem nerve cells, the nucleus had two or even three of these immunoreactive NI (Fig. 9.2) (Rüb et al. 2014a). *Intra-axonal aggregations immunoreactive for the p62 protein*, likewise, turned out to be a consistent and widespread feature of the brainstem pathology of HD (Figs. 9.3, 9.4, and 9.5). They were ubiquitously present in all brainstem fiber tracts of HD patients (i.e., pyramidal tract, all cerebellar peduncles, medial and lateral lemnisci, medial longitudinal fascicle, trapezoid body, pontocerebellar and olivocerebellar fibers, pontine reticular formation, spinocerebellar tracts, cuneate and gracile fascicles), irrespective of the length, caliber, or myelinization of the affected axons (Rüb et al. 2014a). Intra-axonal inclusions were present in the axon hillock or in more distal parts of the axon and in some instances occurred at multiple sites of an affected axon. Their shape varied from round, ovoid, club shaped, irregular, ramified, and torpedo-like, and their diameter was identical to that of the affected axon or larger than the normal axonal diameter (Fig. 9.3). The composition of the intra-axonal inclusions in the brainstem fiber tracts shared many biological similarities with that of NI. They could also be immunolabeled with the anti-polyglutamine antibody 1C2, and, along with the proteasomal shuttle protein p62 and the pathologically altered form of the disease protein Htt, these intra-axonal inclusions also contained ubiquitin (Figs. 9.4 and 9.5) (Rüb et al. 2014a).

Although previous neuropathological studies contributed to a better understanding of the biochemical composition and the distribution of NI in the brains of HD patients, the pathophysiological mechanisms of their formation and pathogenetic relevance for the neurodegenerative process of HD are still uncertain (Borrell-Pagès et al. 2006; DiFiglia et al. 1997; Gil and Rego 2008; Gourfinkel-An et al. 1998; Gutekunst et al. 1999; Imarisio et al. 2008; Kuemmerle et al. 1999; Maat-Schieman et al. 1999; Ortega et al. 2007; Sapp et al. 1999; Sieradzan et al. 1999; Valera et al. 2005; Wooten et al. 2006). Based on the results of experimental studies, but apparently highly anthropomorphical and mechanistical, toxic or protective effects directly responsible for the fate of affected nerve cells have been assigned to these NI by many researchers in the field (Borrell-Pagès et al. 2006; Imarisio et al. 2008; Kuemmerle et al. 1999; Li and Conforti 2013; Li and Li 2011; Maat-Schieman et al. 1999; Ortega et al. 2007). In any case, light microscopically visible NI are aberrant components of nerve cells which develop on the basis of the pathological process of HD. They represent a prominent, widespread, and established immunocytochemical morphological feature of the HD-associated brain pathology and serve as an important neuropathological diagnostic marker of HD. In spite of their pathognomonic features, currently the possible mechanisms remain unclear through which NI may ultimately contribute to neuronal survival or damage of affected nerve cells. Accordingly, their exact role for the fate of affected nerve cells in HD is still uncertain. Nevertheless, the presence of NI in degenerated and well-preserved brainstem nuclei of HD patients underscores a striking discrepancy, but no correlation between the distribution pattern of NI and that of neurodegeneration (Borrell-Pagès et al. 2006; Gunawardena and Goldstein 2005;

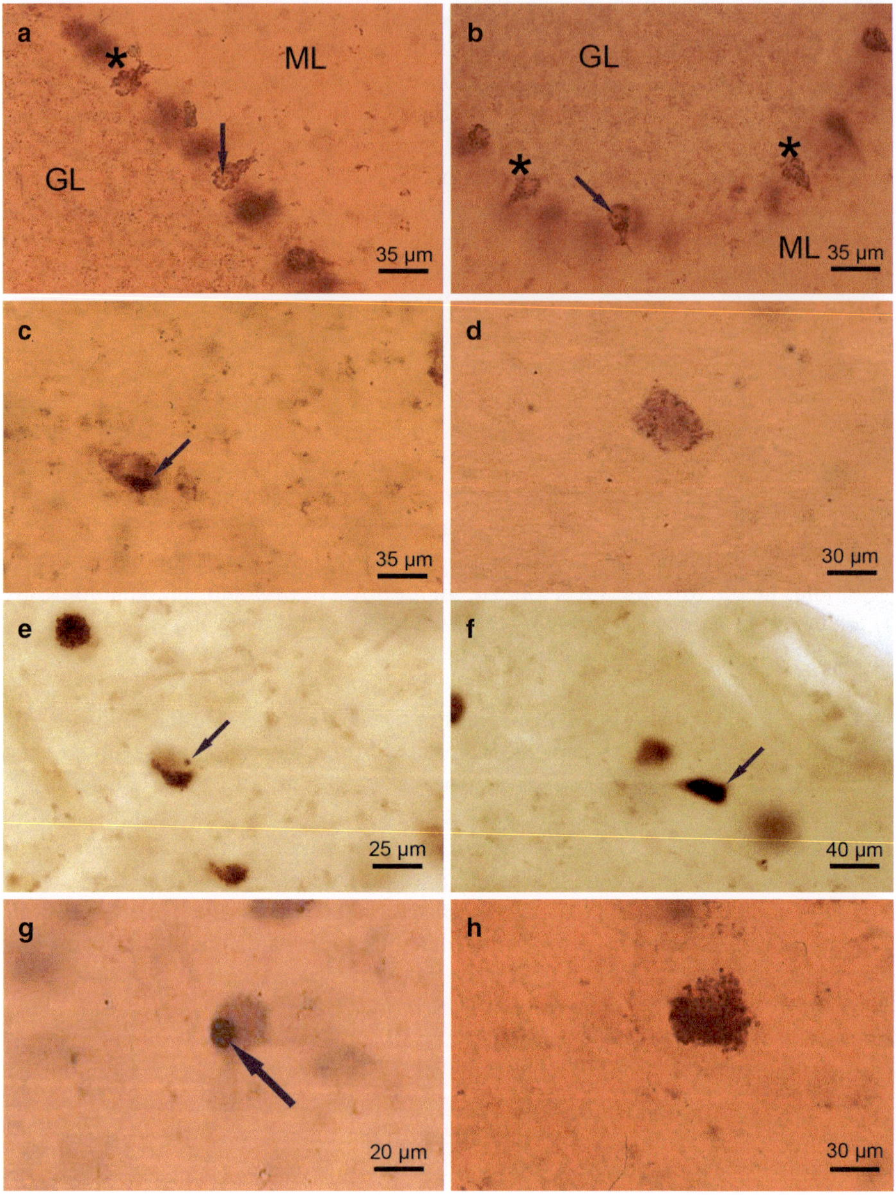
a
ML
GL
35 µm
b
GL
ML
35 µm
c
35 µm
d
30 µm
e
25 µm
f
40 µm
g
20 µm
h
30 µm

Gutekunst et al. 1999; Imarisio et al. 2008; Maat-Schieman et al. 1999; Rüb et al. 2014a; Sapp et al. 1999; Sieradzan et al. 1999). This striking discrepancy and the lack of correlation (1) do not support an immediate toxic or protective role of NI, (2) favor the view that these intranuclear protein aggregates are not directly linked to nerve cell death or survival and (3) support the assumption that NI are not directly decisive for the destiny of affected nerve cells in HD. Since they yet represent aberrant constituents of brain nerve cells, NI still can serve and should be regarded further on as *pathognomonic and reliable morphological biomarkers* facilitating the neuropathological diagnosis of HD (Atkin and Paulson 2014; Becher et al. 1998; Borrell-Pagès et al. 2006; DiFiglia et al. 1997; Finkbeiner and Mitra 2008; Gil and Rego 2008; Gourfinkel-An et al. 1998; Gunawardena and Goldstein 2005; Gutekunst et al. 1999; Imarisio et al. 2008; Kuemmerle et al. 1999; Labbadia and Morimoto 2013; Li and Conforti 2013; Li and Li 2011; Maat-Schieman et al. 1999, 2007; Ortega et al. 2007; Sapp et al. 1999; Schulte and Littleton 2011; Sieradzan et al. 1999; Valera et al. 2005; Walker 2007b).

The results of recent experimental studies (1) suggested various direct and/or indirect roles of the disease protein Htt for axonal transport processes (see Sect. 8.1) and (2) pointed to the occurrence of *axonal transport deficits* as a significant factor that

Fig. 9.6 Neuronal protein inclusions in the surviving nerve cells of the cerebellum in Huntington's disease (HD). (**a**, **b**) Remaining *Purkinje cells* in the vermal portion of the cerebellum of a Huntington's disease (HD) patient with Vonsattel grade 2 of striatal atrophy. Fine and granular dispersed cytoplasmic protein aggregates (*asterisks*), as well as dot-like neuronal intranuclear inclusions (*arrows*). (**c**) Surviving nerve cell of the cerebellar *fastigial nucleus* (FN) of an HD patient with Vonsattel grade 2 of striatal atrophy showing a large compact cytoplasmic protein aggregate (*arrow*). (**d**) Remaining FN nerve cell of a typical HD patient with Vonsattel grade 3 of striatal atrophy depicting granular cytoplasmic protein aggregates. (**e**) Remaining nerve cell of the cerebellar *emboliform nucleus* (e) of an HD patient with Vonsattel grade 3 of striatal atrophy with an intranuclear inclusion (*arrow*) and granular cytoplasmic aggregates. (**f**) Remaining E nerve cell of the same HD patient with a compact cytoplasmic inclusion (*arrow*). (**g**) Remaining nerve cell of the cerebellar *dentate nucleus* (DN) of an HD patient with Vonsattel grade 2 of neostriatal atrophy with a compact cytoplasmic inclusion (*arrow*). (**h**) Surviving DN nerve cell of the same HD patient with granular cytoplasmic inclusions. (**a–h** 1C2 immunostaining; 100 μm polyethylene glycol sections) (Modified according to Rüb et al. (2013a) (Figure 2, page 170; Figure 4, page 172; Figure 5, page 173; Figure 6, page 174); with kind permission from John Wiley and Sons) Abbreviations: *GL* granular layer of the cerebellum, *ML* molecular layer of the cerebellum contributes to the damage and death of nerve cells in HD (Fig. 8.2) (Borrell-Pagès et al. 2006; Davies et al. 2007; De Vos et al. 2008; Gunawardena and Goldstein 2005; Li and Conforti 2013; Millecamps and Julien 2013; Walker 2007a). Although impairments of different physiological and biochemical mechanisms may underlie axonal transport deficits in HD, a most likely key factor contributing to intra-axonal transport deficits is a *physical block caused by intra-axonal protein aggregation formation* (Borrell-Pagès et al. 2006; Davies et al. 2007; Gunawardena and Goldstein 2005; Imarisio et al. 2008; Li and Conforti 2013; Millecamps and Julien 2013; Rüb et al. 2014a). Intra-axonal protein aggregations have been previously described in the neostriatum, cerebral neocortex, internal capsule, and cerebral peduncle of HD patients. They occur as torpedo-like inclusions in the remaining cerebellar Purkinje cells of HD patients and may also develop in the cerebellar white matter. In addition, they represent a widespread and consistent feature of the brainstem pathology of HD patients and have been shown to occur in nearly all brainstem fiber tracts of HD patients, irrespective of the length, caliber, or myelinization of the affected axons (Atkin and Paulson 2014; Becher et al. 1998; Borrell-Pagès et al. 2006; DiFiglia et al. 1997; Finkbeiner and Mitra 2008; Gil and Rego 2008; Gourfinkel-An et al. 1998; Gunawardena and Goldstein 2005; Gutekunst et al. 1999; Imarisio et al. 2008; Kuemmerle et al. 1999; Labbadia and Morimoto 2013; Li and Conforti 2013; Li and Li 2011; Maat-Schieman et al. 2007; Ortega et al. 2007; Paul 2008; Sapp et al. 1999; Schulte and Littleton 2011; Sieradzan et al. 1999; Valera et al. 2005; Walker 2007b; Wooten et al. 2006). The formation of intra-axonal aggregations represents an important step in the neurodegenerative process of HD. In view of their size and insolubility, the intra-axonal aggregations are most likely detrimental to normal transport functions inside affected brain fiber tracts. Via physical blockade, they may lead to the perturbation of anterograde and retrograde axonal transport processes. Such a physical blockade alone or in combination with the biochemical impairments ultimately may lead to functional relevant disturbance, failure, and even disintegration of the essential transport mechanisms. The consequences of this *failure and disintegration of intra-axonal trafficking mechanisms* for affected nerve cells are most likely complex and include (1) impaired intra-axonal movement of protein complexes, synaptic vesicles, and neuronal organelles, (2) structural changes in axons and accumulations of vesicles and organelles (e.g., mitochondria), (3) synaptic and mitochondrial displacements and dysfunctions which are detrimental to a sufficient, timely, and locally adjusted ATP production and energy metabolism and result in further dysfunctions of axonal trafficking and ultimately contribute to the demise of affected nerve cells (see Sect. 8.1 and Chap. 11) (Fig. 8.2) (Borrell-Pagès et al. 2006; Davies et al. 2007; De Vos et al. 2008; Finkbeiner and Mitra 2008; Gil and Rego 2008; Gunawardena and Goldstein 2005; Imarisio et al. 2008; Labbadia and Morimoto 2013; Li and Conforti 2013; Li and Li 2011; Millecamps and Julien 2013; Oliveira 2010; Ortega et al. 2007; Pandey et al. 2010; Rüb et al. 2013a, 2014a; Schulte and Littleton 2011; Sieradzan et al. 1999; Valera et al. 2005; Walker 2007a, b).

As in other human neurodegenerative diseases assigned to the *proteinopathies*, *protein-misfolding disorders*, or *protein-conformational disorders*, it has been repeatedly proposed that the normal functions of the *neuronal protein quality control machinery* which subserves as a major defense against short-lived and misfolded proteins in postmitotic nerve cells are compromised and overstretched in HD and that these malfunctions ultimately contribute to intraneuronal protein aggregation, as well as neurodegeneration. Although, the exact mechanisms of the *overload of the neuronal protein quality control machinery* are still under research, they are currently not regarded as a result of a direct inhibition or blockade of the UPS by insoluble oligomers of mutant Htt and/or its intraneuronal aggregations. Instead, the *encroachment of the neuronal protein quality control machinery* is considered to arise from an overwhelming or overload of the UPS by the large amounts of newly and excessively synthesized mutant Htt proteins that cannot be successfully degraded and removed and tend to aggregate into stable, but dangerous intraneuronal inclusions (Atkin and Paulson 2014; Borrell-Pagès et al. 2006; Davies et al. 2007; Finkbeiner and Mitra 2008; Imarisio et al. 2008; Jucker and Walker 2011; Labbadia and Morimoto 2013; Li and Li 2011; Millecamps and Julien 2013; Ortega et al. 2007; Renner and Melki 2014; Schulte and Littleton 2011; Walker 2007a, b; Wooten et al. 2006). Overexpression of the *ubiquitin-interacting and proteasomal shuttle protein p62* has been shown to facilitate the formation of such intraneuronal inclusion bodies. In addition, pharmacological proteasome inhibition, depletion of p62, and/or inhibition of the ubiquitination process of aberrant nerve cell proteins have been demonstrated to enhance accumulation and aggregation of damaged polyubiquitinated proteins (Atkin and Paulson 2014; Borrell-Pagès et al. 2006; Davies et al. 2007; Finkbeiner and Mitra 2008; Imarisio et al. 2008; Labbadia and Morimoto 2013; Li and Li 2011; Ortega et al. 2007; Wooten et al. 2006).

The NI and intra-axonal inclusions in HD apparently share several biochemical similarities. The positive immunostaining of intra-axonal inclusions with the antibody 1C2 directed against neuronal proteins with an expanded polyglutamine sequence of more than 37 CAG repeats strongly indicates that these intraneuronal aggregations indeed harbor the mutant disease protein Htt with its elongated polyglutamine tract (Rüb et al. 2014a; Seidel et al. 2010; Trottier et al. 1995). The presence of the *degradation-related proteins ubiquitin and p62* in immunoreactive NI and intra-axonal protein inclusion bodies of HD patients indicates that the *UPS* (1) is indeed involved in the intracellular trafficking of the conformationally changed and misfolded polyglutamine protein Htt, (2) plays a crucial pathogenetical role for the intraneuronal handling and is responsible for the ultimate fate of mutant HTT, and (3) points to defense mechanisms in affected nerve cells which primarily attempt to reduce levels of mutant Htt by its tagging with polyubiquitin chains and subsequent degradation of the polyubiquitin-tagged substrate Htt by the proteolytic multimeric 26S proteasome complex into small recyclable peptides (see Sects. 8.2 and 8.3) (Atkin and Paulson 2014; Borrell-Pagès et al. 2006; Brundin et al. 2010; Davies et al. 2007; Finkbeiner and Mitra 2008; Imarisio et al. 2008; Kuusisto et al. 2008; Kuusisto et al. 2001; Labbadia and Morimoto 2013; Leak 2014; Ortega et al. 2007; Paul 2008; Renner and Melki 2014; Rüb et al. 2014a; Schulte and Littleton 2011; Seibenhener et al. 2004; Walker 2007a; Wooten et al. 2006).

The *co-localization of ubiquitin and p62* within the HD-associated intraneuronal inclusion bodies strongly suggests that the degradation processes in the UPS which are initiated by polyubiquitin tagging of Htt and the interaction of the proteasomal shuttle protein p62 with polyubiquitin chains are interrupted and compromised in HD. The failed attempt of the UPS to remove and degrade the polyubiquitin-tagged Htt may ultimately reflect a decreased or *compromised activity of the UPS or its overload* in HD and lead to the formation of intraneuronal aggregates of the conformationally changed and misfolded Htt (see Sect. 8.2) (Atkin and Paulson 2014; Davies et al. 2007; Finkbeiner and Mitra 2008; Imarisio et al. 2008; Labbadia and Morimoto 2013; Renner and Melki 2014; Schulte and Littleton 2011; Walker 2007a, b).

Since the triggers, sequences, and cascades of the pathogenic events that precede dysfunctions, destabilization, and death of nerve cells in HD are not completely understood, causal or disease-modifying treatments for HD patients are still lacking, unfortunately (Borrell-Pagès et al. 2006; Finkbeiner and Mitra 2008; Imarisio et al. 2008; Labbadia and Morimoto 2013; Ortega et al. 2007; Rüb et al. 2014a; Schapira et al. 2014; Walker 2007a, b). Based on the discovery that multiple mechanisms of the neuronal protein quality control machinery may be impaired in HD, the *main strategies for effective therapeutic intervention*, along with other approaches (e.g., application of targeted small interfering RNA to reduce the large amounts of mutant Htt in the brains of HD patients) are directed toward the recovery of impaired UPS mechanisms (Borrell-Pagès et al. 2006; Davies et al. 2007; Finkbeiner and Mitra 2008; Imarisio et al. 2008; Labbadia and Morimoto 2013; Li and Li 2011; Ortega et al. 2007; Rüb et al. 2014a; Schapira et al. 2014). Although improvement of the UPS activities in affected nerve cells with drugs is technically and practically challenging, it is at present considered as a promising therapeutical strategy (1) to reduce the pathological levels of the polyglutamine protein Htt by increasing its clearance and (2) to prevent the pathological consequences associated with its aggregation in diseased nerve cells (Borrell-Pagès et al. 2006; DiFiglia et al. 1997; Finkbeiner and Mitra 2008; Imarisio et al. 2008; Labbadia and Morimoto 2013; Li and Li 2011; Ortega et al. 2007; Schapira et al. 2014).

According to our current knowledge, multiple types of intraneuronal protein aggregations can develop at different brain sites during the disease process of HD. In addition to the long known intranuclear and intra-axonal aggregations, two other types of cytoplasmic inclusions have been identified in HD: granular and compact forms. Up to now, these *granular and compact cytoplasmic neuronal protein aggregations* have only been described in degenerated brain regions of HD patients (e.g., neocortex, striatum, and cerebellum) (Fig. 9.6) (Gutekunst et al. 1999; Kuemmerle et al. 1999; Rüb et al. 2013a; Sieradzan et al. 1999). Along with NI and intra-axonal aggregations, these types of intraneuronal protein aggregates also represent aberrant constituents of nerve cells that are produced by and formed during the pathological process of HD. In contrast to NI and intra-axonal protein aggregations, almost nothing is known about (1) the conditions, context, and manner of the development of cytoplasmic neuronal protein aggregations, (2) the temporal and spatial relationships of their formation with that of the longer known NI and intra-axonal aggregates, or (3) their exact brain distribution and possible pathogenetic relevance. According to current holistic considerations, the formation of different types of intraneuronal aggregations in different compartments of diseased nerve cells in HD is dynamically regulated by temporarily, spatially, and structurally connected events and passes through intermediate stages (Labbadia and Morimoto 2013; Ortega et al. 2007). Considering this idea of a dynamically regulated process, the formation of neuronal inclusion bodies in different compartments of affected nerve cells should not follow a random process. The different types of intraneuronal inclusions may rather develop in a highly regulated, dynamic, temporal, and spatial pattern, which, however, is incompletely understood and therefore still under research.

Based on our long-lasting experience with the neuronal protein aggregation pathology in the related polyglutamine disease spinocerebellar ataxia type 3 (SCA3), we could yet envisage that the granular and compact cytoplasmic protein aggregations in the nerve cells of HD patients are (1) among the initial and transient events of the intraneuronal protein aggregation process and (2) closely linked to the formation of NI and intra-axonal protein aggregations (Hoche et al. 2008; Rüb et al. 2001, 2008a, 2013b; Seidel et al. 2012).

This chapter was written by Udo Rüb.

Chapter 10
Pathological Nerve Cell Alterations in Huntington's Disease (HD) and Their Possible Role for the Demise of Nerve Cells

10.1 Morphological Alterations of Vulnerable Nerve Cells in Huntington's Disease (HD)

Except for the atrophic neurons in the striatum referred to as neostriatal dark neurons (NDN) (see Sect. 2.2), no consistent *pathognomonic alterations of vulnerable brain nerve cells* have been reported during more than hundred years of morphological HD research. In particular, despite recent careful and systematical investigations of possibly affected brain regions, no nerve cells have been observed in HD brains which display morphological features associated with the occurrence of *classical apoptosis* (i.e., chromatin condensation, nuclear fragmentation, apoptotic bodies) (Graeber and Moran 2002; Rüb et al. 2013a). Classical apoptosis is regarded as a specific form of *programmed cell death* and is defined as a series of stereotypical, biochemical, and morphological alterations leading to nerve cell demise. Classical apoptosis regulates the balance between proliferation and differentiation in the course of brain development and during optimization of adult nerve cell functions. Although it has been implicated in the pathogenesis of neurodegenerative diseases, classical apoptosis cannot account for all cell death phenotypes occurring in these diseases. Therefore, other forms of nerve cell death mechanisms may occur in human diseases affecting the terminally differentiated, postmitotic nerve cells. The absence of apoptotic nerve cells in HD brains supports this point of view and conforms to recent HD studies suggesting that the molecular mechanisms of classical apoptosis do not play a significant role for neurodegeneration occurring in HD patients or in transgenic HD models (Gil and Rego 2008; Graeber and Moran 2002; Pattison et al. 2006; Rüb et al. 2013a).

U. Rüb et al., *The Neuropathology of Huntington's Disease: Classical Findings, Recent Developments and Correlation to Functional Neuroanatomy*,
Advances in Anatomy, Embryology and Cell Biology, Vol. 217,
DOI 10.1007/978-3-319-19285-7_10

10.2 Detection of Ballooned or Chromatolytic Nerve Cells

The observation and first description of so-called *ballooned or chromatolytic nerve cells* in the affected cerebellum of HD patients has changed this situation and may provide an important gateway to better understand the pathogenesis of nerve cell death in HD (Rüb et al. 2013a). Along with shrunken nerve cells, ballooned nerve cells represent the vast majority of remaining cerebellar nerve cells of HD patients and have also been abundantly observed in the degenerated thalamus and brainstem of clinically diagnosed and genetically confirmed HD patients (Figs. 5.3, 5.4, and 5.5) (Rüb et al. 2015; Rüb et al. 2013a) (see Sect. 5.2). Ballooned nerve cells cannot be regarded as disease-specific or pathognomonic neuropathological features of HD. They may also occur in large quantities at multiple brain sites in human tauopathies (i.e., Alzheimer's disease, AD; argyrophilic grains disease, AGD; corticobasal degeneration, CBD; progressive supranuclear palsy, PSP), but their brain distribution pattern in these human tauopathies is different from that observable in HD (Dickson 1999; Fujino et al. 2004; Grinberg and Heinsen 2009; Kato et al. 1992; Rüb et al. 2013a; Stover and Watts 2001). The rounded and massively enlarged cytoplasm of ballooned neurons harbors a homogeneous central basophilic substance. Their Nissl substance, lipofuscin granules, and flattened nucleus are displaced along the cytoplasmic membrane and mimic central chromatolysis. This massive enlargement of the perikarya of ballooned neurons is linked to structural damage of their axons and/or *functional impairment of intra-axonal transport mechanisms* (see Sect. 5.2) (Figs. 5.3, 5.4, and 5.5) (Fujino et al. 2004; Kato et al. 1992; Rüb et al. 2013a).

Recent immunocytochemical studies have shown that *axonal pathologies* in the brains of HD patients along with the occurrence of calbindin-immunoreactive torpedo-like inclusions in the axons of cerebellar Purkinje cells also include the presence of axonal aggregates of the disease protein huntingtin (Htt) in additional gray and white matter brain components of HD patients (e.g., neostriatum, cerebral neocortex, internal capsule, cerebral peduncle), as well as axonal inclusions in brainstem fiber tracts (see Sects. 5.2 and 9.2) (Figs. 5.2, 9.3, 9.4, and 9.5) (Atkin and Paulson 2014; Becher et al. 1998; Borrell-Pagès et al. 2006; Finkbeiner and Mitra 2008; Gil and Rego 2008; Gourfinkel-An et al. 1998; Gunawardena and Goldstein 2005; Gutekunst et al. 1999; Imarisio et al. 2008; Kuemmerle et al. 1999; Labbadia and Morimoto 2013; Li and Conforti 2013; Li and Li 2011; Maat-Schieman et al. 2007; Ortega et al. 2007; Paul 2008; Rüb et al. 2014a; Schulte and Littleton 2011; Sieradzan et al. 1999; Valera et al. 2005; Walker 2007b; Wooten et al. 2006). Considering the occurrence of these diverse axonal pathologies in the brains of HD patients and the well-known association between axonal damage and the transition of affected nerve cells toward ballooned nerve cells, it appears conceivable that elucidation of the significance and role of the HD-associated axonal pathology in the pathogenesis of ballooned nerve cells (1) may represent a new gateway to the pathological process of HD, (2) can provide new and important insights into the pathological mechanisms of nerve cell demise in HD, and (3) may lead to the identification of early and crucial steps in the cascades of pathological events that via the intermediate stage of nerve cell swelling eventually lead to destabilization, shrinkage, and death of susceptible nerve cells. These cascades of *pathological events preceding demise of nerve cells* may include the following: (1) An initial and yet unknown causative pathological condition or event in diseased nerve cells that initiates the pathological cascades and sooner or later leads to axonal dysfunctions and/or damage, swelling or ballooning of affected nerve cells, and displacement and compression of their cellular organelles. (2) Loss of the balance between pathological and protective mechanisms (i.e., expression of heat shock proteins, HSP) competing and interfering in these vulnerable nerve cells. (3) Failure and exhaustion of protective mechanisms in ballooned nerve cells. (4) Loss of nerve cell organelles (i.e., mitochondria, Nissl substance) and severe shrinkage of the nucleus and cytoplasm of nerve cells. (5) Ultimate degeneration and demise of exhausted nerve cells leaving behind structurally stable lipofuscin granules in the neuropil which represent reliable markers of the former position of demised nerve cells (Braak et al. 2003b; Rüb et al. 2001, 2003a, 2004a, 2008a, b, 2009, 2013b; Scherzed et al. 2012).

Mitochondria are well known for their role in cellular energy supply via ATP generation through oxidative phosphorylation. Since their cellular functions are also needed throughout the axon of nerve cells, the functional integrity, unimpaired intracellular trafficking, as well as positioning of mitochondria in accordance to local energy requirements are therefore pivotal for the survival of nerve cells. The traditional view of mitochondria of nerve cells as powerhouses quietly lingering around and resting in the cytosol of cells has been stepwise replaced in the past by the perspective of a dynamic intraneuronal mitochondrial trafficking network that physically connects remote cellular compartments (such as neuronal synapses) to the soma of nerve cells and is thus involved in major cellular functions. After biogenesis in the cell body of nerve cells, mitochondria together with other cellular organelles and newly synthesized protein complexes are delivered into the axon, loaded onto molecular kinesin motors, and anterogradely transported along microtubule tracks to the synaptic terminal (Fig. 8.2) (De Vos et al. 2008; Gunawardena and Goldstein 2005; Li and Conforti 2013; Millecamps and Julien 2013; Ortega et al. 2007). While some of the anterogradely transported mitochondria move persistently over long distances (tens to hundreds of micrometers), others appear anchored or remain stationary in the axon according to specific local energy demands. Since accumulating oxidative damage impairs the functions of mitochondria during the period of their service in the axon, aged and impaired mitochondria are returned via molecular dynein motors and retrograde transport mechanisms to the nerve cell body for degradation (see Sect. 8.1) (De Vos et al. 2008; Gunawardena and Goldstein 2005; Li and Conforti 2013; Millecamps and Julien 2013; Ortega et al. 2007).

The axonal pathologies observed in the brains of HD patients most likely are detrimental to *axonal transport mechanisms*, which in turn result in disruption of adequate *mitochondrial trafficking along axons* and deficient neuronal energy production and supply (see Sects. 8.1, 9.1, 9.2, and 10.2) (Borrell-Pagès et al. 2006; Davies et al. 2007; De Vos et al. 2008; Finkbeiner and Mitra 2008; Gil and Rego

2008; Gunawardena and Goldstein 2005; Imarisio et al. 2008; Labbadia and Morimoto 2013; Li and Conforti 2013; Li and Li 2011; Millecamps and Julien 2013; Oliveira 2010; Ortega et al. 2007; Pandey et al. 2010; Rüb et al. 2013a, 2014a; Schulte and Littleton 2011; Walker 2007a, b). Since failure of mitochondrial energy production is associated with a variety of late-onset neurodegenerative diseases including HD, the involvement of mitochondria and their dysfunction in the pathogenic context of neurodegenerative disorders are now generally accepted. Although additional impairments may contribute to mitochondrial dysfunction and decreased energy production in HD (Borrell-Pagès et al. 2006; Davies et al. 2007; De Vos et al. 2008; Finkbeiner and Mitra 2008; Gil and Rego 2008; Gunawardena and Goldstein 2005; Imarisio et al. 2008; Labbadia and Morimoto 2013; Li and Conforti 2013; Li and Li 2011; Millecamps and Julien 2013; Oliveira 2010; Ortega et al. 2007; Pandey et al. 2010; Rüb et al. 2013a, 2014a; Schulte and Littleton 2011; Walker 2007a, b), damage to mitochondrial trafficking most likely leads to a displacement of these neuronal powerhouses, is detrimental to a timely ATP production and energy supply in accordance to the local demands of specific nerve cell sites, and thus plays a significant role in the pathogenesis of neuronal dysfunction and death during HD. Such impairments of mitochondrial transport not only provide suitable explanations for the disturbed energy situation described in HD patients, but may also cause an accumulation of these essential organelles in the cell body of affected nerve cells, which provokes the massive enlargement of their cytoplasm ultimately giving shape to ballooned or chromatolytic nerve cells. Furthermore mitochondrial dysfunction, which reduces *ATP production and energy metabolism*, initiates a vicious circle that ultimately leads to further deficits of axonal transport processes and to structural damage of axons (see Sect. 8.1) (Borrell-Pagès et al. 2006; Davies et al. 2007; De Vos et al. 2008; Finkbeiner and Mitra 2008; Gil and Rego 2008; Gunawardena and Goldstein 2005; Imarisio et al. 2008; Labbadia and Morimoto 2013; Li and Conforti 2013; Li and Li 2011; Millecamps and Julien 2013; Oliveira 2010; Ortega et al. 2007; Pandey et al. 2010; Rüb et al. 2013a, 2014a; Schulte and Littleton 2011; Walker 2007a, b).

In search for the underlying pathophysiological mechanisms leading to nerve cell death in HD, it should not be overlooked that in ballooned neurons occurring in other human neurodegenerative diseases, *cytoplasmic HSP* are upregulated. In these diseases and their experimental models, HSP are thought to serve as a defending principle and as potent suppressors of neurodegeneration (Atkin and Paulson 2014; Borrell-Pagès et al. 2006; Finkbeiner and Mitra 2008; Fujino et al. 2004; Imarisio et al. 2008; Labbadia and Morimoto 2013; Leak 2014; Li and Li 2011; Ortega et al. 2007; Paul 2008; Rüb et al. 2013a; Wooten et al. 2006). Restoration or enhancement of protein folding capacity through upregulation of molecular chaperones has been demonstrated in experimental disease models to enhance mutant Htt degradation and suppression of its tendency to form intraneuronal aggregations. Accordingly, *upregulation of cytoplasmic HSP in ballooned or chromatolytic nerve cells* may also act as a neuroprotective mechanism in HD and could also be regarded as a promising therapeutic target for the polyglutamine "protein misfolding disorder" HD (Atkin and Paulson 2014; Borrell-Pagès et al. 2006; Brundin et al. 2010;

Finkbeiner and Mitra 2008; Fujino et al. 2004; Goedert et al. 2010; Imarisio et al. 2008; Jucker and Walker 2011; Labbadia and Morimoto 2013; Leak 2014; Ortega et al. 2007; Pattison et al. 2006; Renner and Melki 2014; Rüb et al. 2013a; Wooten et al. 2006). Since improving molecular chaperone defenses can help to reduce the detrimental effects of the defective disease protein Htt and associated neurodegeneration in HD, the role and relevance of neuronal HSP in the surviving ballooned nerve cells of HD patients should be examined more carefully in future studies.

This chapter was written by Udo Rüb and Horst-Werner Korf.

Chapter 11
Conclusions and Outlook

The historical survey of the evolution of the knowledge in neuropathological HD research provided in this monograph shows that the stepwise scientific progress made during a time period of more than one century (1) has considerably changed the *traditional, reductionistic pathoanatomical, and pathophysiological concepts of the polyglutamine disease HD*, which were unilaterally based on the well-known degeneration of the striatum and (2) have paved the way for basic research aimed at elucidation of the pathogenic mechanisms leading to HD. The focused efforts and progress in neuropathological HD research led to the development, establishment, and international appreciation of a simple, but reliable grading system of the chronological and topographical progression and severity of the HD-related degeneration in the striatum. They form the empirical base for the implementation of the cerebral allo- and neocortex as additional main targets of the disease process of HD (Figs. 1.4, 2.1, 2.2, 2.4, 2.5, 2.6, 2.7, 2.8, 3.1, 3.2, and 3.3) and favor the view that the consistent degeneration of the pallidum, select thalamic nuclei, cerebellum, and brainstem also represents inherent features of the HD brain pathology (Figs. 2.10, 2.11, 4.4, 4.5, 5.2, 5.3, 5.4, 5.5, 6.4, and 6.5) (see Chaps. 1, 2, and 3) (Atkin and Paulson 2014; Borrell-Pagès et al. 2006; Braak and Braak 1992a, b; Bruyn et al. 1979; De la Monte et al. 1988; Estrada-Sanchez and Rebec 2013; Fennema-Notestine et al. 2004; Ferrante et al. 1987; Finkbeiner and Mitra 2008; Hedreen et al. 1991; Heinsen et al. 1992, 1994, 1996, 1999; Heinsen and Rüb 1997; Imarisio et al. 2008; Lange 1981; Lange and Aulich 1986; Lange et al. 1976; Li and Conforti 2013; Margolis and Ross 2003; Myers et al. 1988; Rosas et al. 2003; Rüb et al. 2009, 2013a, 2014a, b; Schulte and Littleton 2011; Selemon et al. 2004; Sotrel et al. 1991; Valera et al. 2005; Vonsattel 2008; Vonsattel and DiFiglia 1998; Vonsattel et al. 1985; Walker 2007a, b).

By showing that the extent of brain degeneration has been considerably underestimated, these studies have helped to unravel the real extent of brain neurodegeneration in HD and showed (1) that brain neurodegeneration in HD is more widespread

U. Rüb et al., *The Neuropathology of Huntington's Disease: Classical Findings, Recent Developments and Correlation to Functional Neuroanatomy*,
Advances in Anatomy, Embryology and Cell Biology, Vol. 217,
DOI 10.1007/978-3-319-19285-7_11

and severe than reported in previous studies, (2) that the distribution pattern of neuronal loss in the brain of HD patients goes far beyond the well-known subcortical predilection site of the underlying disease process (i.e., striatum) and its targets in the cerebral cortex, (3) and that the distribution pattern of HD brain neurodegeneration has more neuropathological similarities and displays far more overlap with the brain lesional pattern described in the *related polyglutamine spinocerebellar ataxias types 1 (SCA1), 2 (SCA2), and 3 (SCA3)* than previously thought (Hoche et al. 2008; Lastres-Becker et al. 2008; Riess et al. 2008; Rüb et al. 2003a, 2004a, b, 2005, 2008a, b, 2013b, 2014b; Scherzed et al. 2012; Seidel et al. 2012). Thus, the traditional reductionistic view of HD as a primary and exclusive disease of the striatum has been challenged and is outdated in the meantime: the new findings of neurodegeneration at multiple cortical and subcortical brain sites ultimately form the empirical base of the currently favored concept of the polyglutamine disease HD as a *polytopic or multisystem degenerative disease* of the human brain (Heinsen et al. 1994; Lange and Aulich 1986; Rüb et al. 2013a; Vonsattel 2008). Further results of postmortem studies in support of the multisystem character of HD suggested that a number of additional subcortical regions (e.g., amygdala, hypothalamus, subthalamic nucleus, claustrum), which are intimately linked with the well-known targets of the disease process of HD via fiber tracts (e.g., prefrontal cortex, entorhinal and transentorhinal regions, thalamic mediodorsal nucleus and centromedian-parafascicular complex, striatum), may also undergo neurodegeneration during HD (see Chap. 1) (Fig. 1.4) (Lange et al. 1976; Lange and Aulich 1986; Van Wamelen et al. 2014; Vonsattel 2008; Vonsattel and DiFiglia 1998; Vonsattel et al. 1985). However, according to our current experience, the involvement of these brain grays may be difficult to be assessed, and it may also represent an inconsistent feature of HD that depends on the disease duration or is actually characterized by a subregional distribution of neurodegeneration that contradicts the current neuropathological literature. Therefore, further confirmative postmortem studies of these brain sites in HD are recommended.

In the *most prevalent human synucleinopathy* (i.e., *Parkinson's disease*, *PD*), the motor and nonmotor disease symptoms have been exclusively and unilaterally explained by the well-known degeneration of the compact part of the substantia nigra for a long time. However, this traditional view has been revised during the last three decades. Since then we have learned much about the degenerative process, the different types of PD-related neuronal and glial protein aggregations, and the brain extent, distribution, and propagation of the PD-related aggregation pathology and coexistent neuronal loss. We have obtained a refined picture of the neuropathology of this neuronal and oligodendroglial synucleinopathy which led to the replacement of previous concepts by the empirically based consideration of *PD as a multisystem disorder of the human brain*. According to the improved neuropathological knowledge, the PD-related pathology involves not only the midbrain mesostriatal dopaminergic system but also the mesolimbic dopaminergic projection system, the cholinergic systems of the basal forebrain and midbrain, the histaminergic hypothalamus, as well as the noradrenergic and serotonergic brainstem systems. The disease process of PD ultimately results in a brain pathology that is widely distributed over interconnected gray components of the cerebral cortex, basal forebrain, diencephalon, and brainstem and characteristically leads to a severe involvement of the cortical and subcortical components of the limbic system of the brain. This improved pathoanatomical knowledge facilitated advanced and more precise clinicopathological correlations in PD and offered new and more adequate morphological explanations for a large spectrum of motor and nonmotor PD symptoms that have been previously attributed to the well-known damage to the dopaminergic substantia nigra (e.g., vestibular, ingestive, and oculomotor dysfunctions) (Braak et al. 2003a, 2003c, 2004).

Similar to the situation in the synucleinopathy PD, many disease symptoms of the *polyglutamine disease HD* including somatomotor, oculomotor, cognitive, and psychiatric dysfunctions have been more or less mechanically and exclusively explained by the functional consequences of the prominent neuronal loss in the striatum, the subcortical predilection site of the HD-related pathology which for a long time has been regarded as the primary and only target of the underlying disease process (Mink 1996; Walker 2007a). However, the identification and description of neurodegeneration in a number of extrastriatal brain sites of HD patients resulted in an improved pathoanatomical HD knowledge. This improved knowledge, in turn, provided new insights into the morphological basis of a large spectrum of HD-related disease symptoms and, likewise, offered for the first time appropriate explanations for disease signs that were previously associated with neostriatal damage (e.g., oculomotor dysfunctions, dysphagia, incoordination, falls, ataxia, imbalance, cognitive decline, psychiatric manifestations).

Despite progress in HD research, further pathoanatomical brain studies are required to elucidate the possible morphological counterparts of some less well-explained disease symptoms of HD patients (e.g., weight loss, dysfunctions of vertical saccades, vergence, and steady fixation). In the advanced clinical stages, most affected HD patients suffer from an *unintended and severe weight loss* that most commonly leads to cachexia, which represents a frequent cause of death in HD. This tremendous weight loss of HD patients for a long time has been unilaterally explained by an additional energy expenditure caused by the excessive occurrence of unwanted choreatic movements. Most recent studies, however, have shown that weight loss in HD by no means is a secondary effect of hyperkinetic symptoms or is associated with a reduced food or caloric intake. It occurs despite adequate or even increased caloric intake, is particularly marked in the final hypokinetic stages of HD, and may result from an altered intrinsic metabolic functioning which is under control of the hypothalamus (Aziz et al. 2008; Borrell-Pagès et al. 2006; Bruyn et al. 1979; Gil and Rego 2008; Kremer et al. 1992; Petersen et al. 2005; Rüb et al. 2013a, 2014a; Walker 2007a, b). Accordingly, accurate postmortem reexaminations of the pathoanatomy of the hypothalamus of patients in different clinical stages of HD and with different Vonsattel grades of striatal atrophy by means of reliable morphological methods are mandatory.

The midbrain *rostral interstitial nucleus of the medial longitudinal fascicle* (*riMLF*) represents the immediate premotor structure for the performance of vertical and torsional saccades and harbors the premotor burst neurons necessary for the generation of these eye movements (Fig. 6.1) (Büttner-Ennever 2006; Büttner and

Büttner-Ennever 2006; Büttner-Ennever and Horn 1997, 2004; Büttner-Ennever et al. 1982; Horn 2006; Horn and Büttner-Ennever 1998, 1999; Leigh and Kennard 2004; Leigh and Zee 2006; Rüb et al. 2008b, 2009). The more caudally located *superior colliculus* (*SC*) is crucial for the initiation of saccades, contributes to vergence, and is involved in the suppression of unwanted saccades during steady target fixation (Fig. 6.1) (Büttner and Büttner-Ennever 2006; Büttner-Ennever and Horn 1997; Horn 2006; Leigh and Zee 2006; Rüb et al. 2008b, 2009). Since these two midbrain premotor oculomotor nuclei have never been subjected to accurate neuropathological investigations, studies of the riMLF and SC are necessary to improve our knowledge about the pathoanatomical base of dysfunctions of vertical saccades, as well as impairments of vergence and steady target fixation that may occur in HD patients (see Sect. 6.6).

According to our current neuropathological knowledge and evidence-based theories, the underlying pathological process of HD apparently does not develop, propagate, or operate according to the random principle or in an arbitrary fashion. In contrast, this destructive process proceeds with a selective vulnerability; apparently targets only a subset of subcortical nuclei, as well as distinct neuronal layers of specific areas of the cerebral cortex; and ultimately leads to an interindividually constant damage to a subset of cortical and subcortical brain regions. At the present, the reasons why some brain regions are more heavily affected than others during HD are unclear, and the pathophysiological background of the evolution of the highly characteristic, disease-specific, and consistent distribution pattern of neuronal loss in HD is still enigmatic. Despite this, we and other researchers are guided by the idea that the anatomical interconnectivities between affected brain grays and intra-axonal mechanisms may play a central role in the progression of the pathological process of HD. According to this idea that takes the multisystem character of the polyglutamine disease HD into account, the underlying nonrandom disease process of HD might spread transneuronally along anatomical pathways throughout the brain and, in a stepwise manner, might sequentially propagate from the diseased brain components to the next vulnerable brain site in the neural chain as has also been proposed for the tauopathy Alzheimer's disease (AD) and the synucleinopathy PD (Braak et al. 2003c; Brundin et al. 2010; Costanzo and Zurzolo 2013; Duyckaerts et al. 2009; Goedert et al. 2010; Itzhaki et al. 2004; Jucker and Walker 2011; Labbadia and Morimoto 2013; Lace et al. 2007; Millecamps and Julien 2013; Norrby 2011; Renner and Melki 2014). Indeed, the pattern of brain neurodegeneration in HD elaborated during the last decades apparently follows the neuronal fiber tracts described in nonhuman primates by modern tracing studies and is compatible with the pathoanatomical ideas that pointed to the *possible role of intact anatomical interconnectivities* for the spread of the HD-related pathology throughout the brain. The pattern of neurodegeneration suggests (1) that the degenerated cortical and subcortical brain areas are directly anatomically interconnected and (2) that the pathological process of HD may use these anatomical pathways and interconnectivities for its transneuronal spread and propagation throughout the brain (Brundin et al. 2010; Costanzo and Zurzolo 2013; Goedert et al. 2010; Jucker and Walker 2011; Labbadia and Morimoto 2013; Millecamps and Julien 2013; Norrby 2011; Renner and Melki 2014; Rüb et al. 2014b). The mechanisms responsible for the targeted spread and neuron-to-neuron propagation along anatomical pathways throughout the diseased brain may include (1) physical blockade of axonal trafficking by intra-axonal protein aggregates and swellings, (2) inhibition of synaptic functions by impairments of anterograde axonal transport mechanisms, and (3) transsynaptic and interneuronal spread and propagation of the disease process in a prion-like manner. Although the widespread occurrence of intra-axonal immunoreactive protein aggregations in brain fiber tracts of HD patients conforms to the idea that HD in fact may represent a chronically progredient prion-like protein misfolding disease, further studies are required to prove the suggested *prion-like nature of the disease process of HD* (see Sects. 3.5, 4.1, and 10.2) (Brundin et al. 2010; Costanzo and Zurzolo 2013; Goedert et al. 2010; Jucker and Walker 2011; Labbadia and Morimoto 2013; Millecamps and Julien 2013; Norrby 2011; Renner and Melki 2014).

In both, the human tauopathy AD and the human synucleinopathy PD, neuropathological staging procedures have been proposed that capture the different developmental stages of the disease-related pathologies, their proposed possible brain induction sites, as well as their final brain distribution pattern and extent (Braak and Braak 1991; Braak et al. 2003a, 2003c, 2004, 2006; Goedert et al. 2010; Jucker and Walker 2011; Seidel et al. 2015). It would be a seminal task to extend and complement the description of the spatially and chronologically spread of neurodegeneration within the striatum according to the *Vonsattel grading system* by the results of further cross-sectional studies of HD brains and to develop an extended HD staging procedure that takes all aspects of the brain pathology into consideration (see Chap. 2) (Figs. 2.1, 2.2, 2.4, and 2.5) (Vonsattel 2008; Vonsattel and DiFiglia 1998; Vonsattel et al. 1985).

As holds true for AD and PD, the reconstruction of the exact *spatiotemporal propagation of the degenerative process throughout the HD brains* (e.g., by means of comprehensive cross-sectional studies) may help to identify the actual site of origin of the degenerative process in the brain and will show whether the neostriatum is indeed the brain site undergoing initial damage during HD or whether it is only secondarily affected after the cerebral cortex. These future studies (1) will lead to improved insights into the pathophysiological mechanisms of the degenerative process of HD; (2) will help to elucidate the essential principles, mechanisms, and pathways of its topographical and chronological spread and propagation throughout the brain; and (3) will also offer conclusive explanations of the enigmatic phenomenon that only distinct cortical and subcortical brain sites consistently undergo neurodegeneration during HD.

Despite the fact that multiple efforts are under way to change this situation, unfortunately, no established *causative treatment or effective disease-modifying therapy* exists for HD patients which would be able to delay the onset or prevent and reverse the progression of the disease. To date, only *symptomatic treatments* of chorea, dystonia, other movement disorders, and nonmotor aspects of HD can be offered which may have favorable effects on disease symptoms and quality of life of affected patients. The spectrum of symptomatic pharmacological treatment strategies is complemented by a number of non-pharmacological interventions, such as surgical procedures; adjunctive, alternative, and complementary therapies (e.g., physical therapy, speech therapy, occupational therapy, exercise therapy, music therapy, dance, or video game playing); behavioral plans; and cognitive interventions, which also may play a role in addressing the symptoms of HD and need also to be considered when choosing medications (Borrell-Pagès et al. 2006; Finkbeiner and Mitra 2008; Imarisio et al. 2008; Labbadia and Morimoto 2013; Ortega et al. 2007; Schapira et al. 2014; Walker 2007a, b).

To effectively and permanently change this unsatisfactory therapeutical situation, there is pressing need for further studies aiming to decipher the enigmatic cascades of pathophysiological and pathobiochemical mechanisms that eventually lead to neurodegeneration in HD. Before their results can ultimately contribute to the future development of efficient causative and/or disease-modifying therapies for HD, the following questions already raised in the preceding chapters of this monograph need to be addressed (see Chaps. 8, 9, and 10):

1. Is the mutated, misfolded, and conformationally changed huntingtin protein (Htt) via its large variety of protein interactions involved in physiological and/or biochemical mechanisms that can directly or indirectly contribute to impairments of axonal transport processes in the brains of HD patients under pathological conditions?
2. Are the intra-axonal protein aggregations that form in brain fiber tracts of HD patients detrimental to normal anterograde and retrograde axonal transport processes?
3. Do intra-axonal protein aggregations impair the directed intra-axonal movement of protein complexes, synaptic vesicles, and neuronal organelles along axons and provoke structural changes in axons?
4. Does damage to intra-axonal mitochondrial trafficking contribute to a reduced ATP production and disturbed energy situation in HD?
5. Is the occurrence of intra-axonal protein aggregations confined to the currently known brain predilection sites (e.g., cerebellum, brainstem), or are they also present in additional brain fiber tracts that establish the anatomical interconnectivities between the cerebral cortex, thalamus, and striatum?
6. Are the protein aggregations in neuronal intranuclear inclusions (NI) merely pathognomonic and reliable morphological markers of HD, or are they responsible for the fate, survival, or death of affected nerve cells?
7. Under which pathological conditions and context do compact and granular cytoplasmic neuronal protein aggregations develop in HD?
8. Is the formation of the compact and granular cytoplasmic neuronal protein aggregations spatially and timely related to the evolution of NI and/or intra-axonal aggregates?
9. How are the compact and granular cytoplasmic neuronal protein aggregations distributed anatomically in the brains of HD patients?
10. Does the presence of the *degradation-related proteins ubiquitin and p62* in immunoreactive NI and intra-axonal protein inclusion bodies actually indicate that the ubiquitin-proteasome pathway (UPS) is crucial for the intraneuronal trafficking and handling of mutant Htt?

11. Is the UPS along with molecular chaperones among the neuronal defense mechanisms of affected nerve cells which attempt to reduce the levels of Htt by polyubiquitin tagging and subsequent proteasomal degradation into small peptides?
12. Is the protein degradation process of the UPS, mediated by the proteasomal shuttle protein p62 and polyubiquitin chains, overloaded and compromised in HD and thus unable to remove and to degrade polyubiquitin-tagged Htt?
13. Does failure of the UPS contribute significantly to the formation of unsoluble intraneuronal aggregates of conformationally changed and misfolded polyglutamine Htt?
14. Are heat shock proteins upregulated in the cytoplasm of ballooned or chromatolytic nerve cells? Do they act as a defense mechanism and help to reduce the detrimental effects of the defective Htt protein? Do they serve as potent suppressors of the neurodegenerative process of HD?
15. Do blockades of intra-axonal transport processes also lead to an impairment of mitochondrial trafficking in nerve cells and result in the accumulation of mitochondria which is detrimental to a sufficient, timely, and locally adjusted energy supply and ultimately contributes to a reduction of intraneuronal ATP production and energy metabolism in HD?
16. Does the development of ballooned or chromatolytic nerve cells in HD take place subsequent to the occurrence of defects of intra-axonal transport mechanisms?

Hopefully, all of these open questions will find their adequate answers by future studies. Adequate answers to these pressing questions will lead to a continuation of the promising avenue of neuropathological HD research, constitute the basis of decisive counteractions against the severely disabling polyglutamine disease HD, and eventually may pave the way for the development of effective, evidence-based, causative, and/or disease-modifying therapies for affected patients.

This chapter was written by Udo Rüb.

References

Alzheimer A (1911) Über die anatomische Grundlage der Huntington'schen Chorea und der choreatischen Bewegung überhaupt. Neurol Centralblatt 30:891–892

Andrew SE, Goldberg YP, Kremer B, Telenius H, Theilmann J, Adam S, Starr A, Squitieri S, Lin B, Kalchman MA, Graham RK, Hayden MR (1993) The relationship between trinucleotide (CAG) repeat length and clinical features of Huntington's disease. Nat Genet 4:398–403

Armstrong E (1990) Evolution of the brain. In: Paxinos G (ed) The human nervous system. Academic Press, San Diego, pp 1–16

Arzberger T, Krampfl K, Leimgruber S, Weindl A (1997) Changes of NMDA receptor subunit (NR1, NR2b) and glutamate transporter (GLT1) mRNA expression in Huntington's disease: an in situ hybridization study. J Neuropathol Exp Neurol 56:440–454

Atkin G, Paulson H (2014) Ubiquitin pathways in neurodegenerative disease. Front Mol Neurosci 7:63

Avanzini G, Girotti F, Caraceni T, Spreafico R (1979) Oculomotor disorders in Huntington's chorea. J Neurol Neurosurg Psychiatry 42:581–589

Aylward EH, Nopoulos PC, Ross CA, Langbehn DR, Pierson RK, Mills JA, Johnson HJ, Magnotta VA, Juhl AR, Paulsen J, PREDICT-HD investigators and Coordinators of Huntington Study Group (2011) Longitudinal change in regional brain volumes in prodromal Huntington disease. J Neurol Neurosurg Psychiatry 82:405–410

Aziz NA, van der Marck MA, Pijl H, Olde Rikkert MG, Bloem BR, Roos RA (2008) Weight loss in neurodegenerative disorders. J Neurol 255:1872–1880

Barbas H, Henion THH, Dermon CR (1991) Diverse thalamic projections to the prefrontal cortex in the rhesus monkey. J Comp Neurol 313:65–94

Bartzokis G, Lu PH, Tishler TA, Fong SM, Oluwadara B, Finn JP, Huang D, Bordelon Y, Mintz J, Perlman S (2007) Myelin breakdown and iron changes in Huntington's disease: pathogenesis and treatment implications. Neurochem Res 32:1655–1664

Bastianelli E (2003) Distribution of calcium-binding proteins in the cerebellum. Cerebellum 2:242–262

Becher MW, Kotzuk JA, Sharp AH, Davies SW, Bates GP, Price DL, Ross CA (1998) Intranuclear neuronal inclusions in Huntington's disease and dentatorubral and pallidoluysian atrophy: correlation between the density of inclusions and IT15 CAG triplet repeat length. Neurobiol Dis 4:387–397

U. Rüb et al., *The Neuropathology of Huntington's Disease: Classical Findings, Recent Developments and Correlation to Functional Neuroanatomy*, Advances in Anatomy, Embryology and Cell Biology, Vol. 217, DOI 10.1007/978-3-319-19285-7

Beenen N, Büttner U, Lange HW (1986) The diagnostic value of eye movement recordings in patients with Huntington's disease and their offspring. Electroencephalogr Clin Neurophysiol 63:119–127

Benninghoff A, Drenckhahn D (eds) (2004) Anatomie: Makroskopische Anatomie, Histologie, Embryologie, Zellbiologie, vol 2, 16th edn. Elsevier GmbH, Urban & Fische, Munich

Bhatia KP, Marsden CD (1994) The behavioural and motor consequences of focal lesions of the basal ganglia in man. Brain 117:859–876

Birnbaum G (1941) Chronisch-progressive Chorea mit Kleinhirnatrophie. Arch Psychiatr Nervenk 114:160–182

Blekher TM, Yee RD, Kirkwood SC, Hake AM, Stout JC, Weaver MR, Foroud TM (2004) Oculomotor control in asymptomatic and recently diagnosed individuals with the genetic marker for Huntington's disease. Vision Res 44:2729–2736

Blekher T, Johnson SA, Marshall J, White K, Hui S, Weaver M, Gray J, Yee R, Stout JC, Beristain X, Wojcieszek J, Foroud T (2006) Saccades in presymptomatic and early stages of Huntington disease. Neurology 67:394–399

Bollen E, Reulen JP, den Heyer JC, van der Kamp W, Roos RA, Buruma OJ (1986) Horizontal and vertical saccadic eye movement abnormalities in Huntington's chorea. J Neurol Sci 74: 11–22

Borrell-Pagès M, Zala D, Humbert S, Saudou F (2006) Huntington's disease: from huntingtin function and dysfunction to therapeutic strategies. Cell Mol Life Sci 63:2642–2660

Braak H, Braak E (1991) Neuropathological stageing of Alzheimer-related changes. Acta Neuropathol 82:239–259

Braak H, Braak E (1992a) The human entorhinal cortex: normal morphology and lamina-specific pathology in various diseases. Neurosci Res 15:6–31

Braak H, Braak E (1992b) Allocortical involvement in Huntington's disease. Neuropathol Appl Neurobiol 18:539–547

Braak H, Del Tredici K, Rüb U, de Vos RA, Jansen Steur EN, Braak E (2003a) Staging of brain pathology related to sporadic Parkinson's disease. Neurobiol Aging 24:197–211

Braak H, Rüb U, Del Tredici K (2003b) Involvement of precerebellar nuclei in multiple system atrophy. Neuropathol Appl Neurobiol 29:60–76

Braak H, Rüb U, Gai WP, Del Tredici K (2003c) Idiopathic Parkinson's disease: possible routes by which vulnerable neuronal types may be subject to neuroinvasion by an unknown pathogen. J Neural Transm 110:517–536

Braak H, Ghebremedhin E, Rüb U, Bratzke H, Del Tredici K (2004) Stages in the development of Parkinson's disease-related pathology. Cell Tissue Res 318:121–134

Braak H, Alafuzoff I, Arzberger T, Kretzschmar H, Del Tredici K (2006) Staging of Alzheimer disease-associated neurofibrillary pathology using paraffin sections and immunocytochemistry. Acta Neuropathol 112:389–404

Brandt J, Folstein SE, Folstein MF (1988) Differential cognitive impairment in Alzheimer's disease and Huntington's disease. Ann Neurol 23:555–556

Brodmann K (1909) Vergleichende Lokalisationslehre der Großhirnrinde. Johann Ambrosius Barth, Leipzig

Brundin P, Melki R, Kopito R (2010) Prion-like transmission of protein aggregates in neurodegenerative diseases. Nat Rev Mol Cell Biol 11:301–307

Bruyn GW (1968) Huntington's chorea. Historical, clinical and laboratory synopsis. In: Vinken PJ, Bruyn GW (eds) Handbook of clinical neurology, vol 6, Diseases of the basal ganglia. North-Holland, Amsterdam, pp 298–378

Bruyn GW, Bots GTAM, Dom R (1979) Huntington's chorea: current neuropathological status. Adv Neurol 23:83–93

Busse ME, Wiles CM, Rosser AE (2009) Mobility and falls in people with Huntington's disease. J Neurol Neurosurg Psychiatry 80:88–90

Butters N, Albert MS, Sax D (1979) Investigations of the memory disorders of patients with Huntington's disease. Adv Neurol 23:203–213

Büttner U, Büttner-Ennever JA (2006) Present concepts of oculomotor organization. Prog Brain Res 151:1–42

Büttner-Ennever JA (2006) The extraocular motor nuclei: organization and functional neuroanatomy. Prog Brain Res 151:95–125

Büttner-Ennever JA, Gerrits NM (2004) Vestibular system. In: Paxinos G, Mai JK (eds) The human nervous system, 2nd edn. Elsevier, San Diego, pp 1213–1240

Büttner-Ennever JA, Horn AKE (1997) Anatomical substrates of oculomotor control. Curr Opin Neurobiol 7:872–879

Büttner-Ennever JA, Horn AKE (2004) Reticular formation: eye movements, gaze, and blinks. In: Paxinos G, Mai JK (eds) The human nervous system, 2nd edn. Elsevier, San Diego, pp 479–510

Büttner-Ennever JA, Büttner U, Cohen B, Baumgartner G (1982) Vertical glaze paralysis and the rostral interstitial nucleus of the medial longitudinal fasciculus. Brain 105:125–149

Büttner-Ennever JA, Cohen B, Pause M, Fries W (1988) Raphe nucleus of the pons containing omnipause neurons of the oculomotor system in the monkey, and its homologue in man. J Comp Neurol 267:307–321

Campbell AM, Corner B, Norman RM, Urich H (1961) The rigid form of Huntington's disease. J Neurol Neurosurg Psychiatry 24:71–77

Carmichael ST, Price JL (1996) Connectional networks within the orbital and medial prefrontal cortex of macaque monkeys. J Comp Neurol 371:179–207

Cattaneo E, Zuccato C, Tartari M (2005) Normal huntingtin function: an alternative approach to Huntington's disease. Nat Rev Neurosci 6:919–939

Coggeshall RE, Lekan HA (1996) Methods for determining numbers of cells and synapses: a case for more uniform standards of review. J Comp Neurol 364:6–15

Collewijn H, Went LN, Tamminga EP, Vegter-Van der Vlis M (1988) Oculomotor defects in patients with Huntington's disease and their offspring. J Neurol Sci 86:307–320

Costanzo M, Zurzolo C (2013) The cell biology of prion-like spread of protein aggregates: mechanisms and implication in neurodegeneration. Biochem J 452:1–17

Cudkowicz M, Kowall NW (1990) Degeneration of pyramidal projection neurons in Huntington's disease cortex. Ann Neurol 27:200–204

Davies JE, Sarkar S, Rubinsztein DC (2007) The ubiquitin proteasome system in Huntington's disease and the spinocerebellar ataxias. BMC Biochem 8(Suppl):1–8

Davison C, Goodhart SP, Shlionsky H (1932) Chronic progressive chorea. The pathogenesis and mechanism; a histopathologic study. Arch Neurol Psychiat 27:906–928

De la Monte SM, Vonsattel JP, Richardson EP Jr (1988) Morphometric demonstration of atrophic changes in the cerebral cortex, white matter, and neostriatum in Huntington's disease. J Neuropathol Exp Neurol 47:516–525

De Vos KJ, Grierson AJ, Ackerley S, Miller CCJ (2008) Role of axonal transport in neurodegenerative diseases. Annu Rev Neurosci 31:1151–1173

Den Dunnen WF (2013) Neuropathological diagnostic considerations in hyperkinetic movement disorders. Front Neurol 4:7

Deuschl G, Lücking CH, Schenck E (1989) Hand muscle reflexes following electrical stimulation in choreatic movement disorders. J Neurol Neurosurg Psychiatr 52:755–762

Dickson DW (1999) Neuropathologic differentiation of progressive supranuclear palsy and corticobasal degeneration. J Neurol 246(Suppl 2):6–15

Difiglia M, Sapp E, Chase KO, Davies SW, Bates GP, Vonsattel JP, Aronin N (1997) Aggregation of huntingtin in neuronal intranuclear inclusions and dystrophic neurites in brain. Science 277:1990–1993

DiProspero NA, Chen EY, Charles V, Plomann M, Kordower JH, Tagle DA (2004) Early changes in Huntington's disease patient brains involve alterations in cytoskeletal and synaptic elements. J Neurocytol 33:517–533

Dom R, Malfroid M, Baro F (1976) Neuropathology of Huntington's chorea. Studies of the ventrobasal complex of the thalamus. Neurology 26:64–68

Dunlap CB (1927) Pathologic changes in Huntington's chorea: with special reference to the corpus striatum. Arch Neurol Psychiatry 18:867–943

Dunlop DS, Mc Hale DM, Lajtha A (1992) Decreased brain N-acetylaspartate in Huntington's disease. Brain Res 580:44–48

Duyao M, Ambrose C, Myers R, Novelletto A, Persichetti F, Frontali M, Folstein S, Ross C, Franz M, Abbott M, Gray J, Coneally P, Young A, Penney J, Hollingsworth Z, Shoulson I, Lazzarini A, Falck A, Koroshetz W, Sax D, Bird E, Vonsattel J, Bonilla E, Alvir J, Bickham Conde J, Cha JH, Dure L, Gomez F, Ramos M, Sanchez-Ramos J, Snodgrass S, de Young M, Wexler N, Moscowitz C, Penchaszadeh G, MacFarlane H, Anderson M, Jenkins B, Srinidhi J, Barnes G, Gusella JMM (1993) Trinucleotide repeat length instability and age of onset in Huntington's disease. Nat Genet 4:387–392

Duyckaerts C, Delatour B, Potier MC (2009) Classification and basic pathology of Alzheimer disease. Acta Neuropathol 118:5–36

Ellenberger C Jr, Petro DJ, Ziegler SB (1978) The visually evoked potential in Huntington disease. Neurology 28:95–97

Eng LF, Ghirnikar RS (1994) GFAP and astrogliosis. Brain Pathol 4:229–237

Eng LF, Ghirnikar RS, Lee YL (2000) Glial fibrillary acidic protein: GFAP-thirty-one years (1969–2000). Neurochem Res 25:1439–1445

Estrada-Sanchez AM, Rebec GV (2013) Role of cerebral cortex in the neuropathology of Huntington's disease. Front Neural Circuits 7:19

Fenelon G, Francois C, Percheron G, Yelnik J (1991) Topographic distribution of the neurons of the central complex (centre median-parafascicular complex) and of other thalamic neurons projecting to the striatum in macaques. Neuroscience 45:495–510

Fennema-Notestine C, Archibald SL, Jacobson MW, Corey-Bloom J, Paulsen JS, Peavy GM, Gamst AC, Hamilton JM, Salmon DR, Jernigan TL (2004) In vivo evidence of cerebellar atrophy and cerebral white matter loss in Huntington disease. Neurology 63:989–995

Ferrante RJ, Kowall NW, Beal MF, Martin JB, Bird ED, Richardson EPJ (1987) Morphologic and histochemical characteristics of a spared subset of striatal neurons in Huntington's disease. J Neuropathol Exp Neurol 46:12–27

Fielding J, Georgiou-Karistianis N, Bradshaw J, Millist L, Churchyard A, Chiu E, White O (2004) Impaired modulation of the vestibulo-ocular reflex in Huntington's disease. Mov Disord 19:68–75

Finkbeiner S, Mitra S (2008) The ubiquitin-proteasome pathway in Huntington's disease. ScientificWorldJournal 8:421–433

Forno LS, Jose C (1973) Huntington's chorea: a pathological study. In: Barbeau A, Chase TN, Paulson GW (eds) Huntington's chorea 1872–1972, vol 1, Huntington's chorea. Raven Press, New York, pp 453–470

François C, Percheron G, Parent A, Sadikot AF, Fenelon G, Yelnik J (1991) Topography of the projection from the central complex of the thalamus to the sensorimotor striatal territory in monkeys. J Comp Neurol 305:17–34

Freeman W, Watts JW (1947) Retrograde degeneration of the thalamus following prefrontal lobotomy. J Comp Neurol 86:65–93

Fujino Y, DeLucia MW, Davies P, Dickson DW (2004) Ballooned neurons in the limbic lobe are associated with Alzheimer type pathology and lack diagnostic specificity. Neuropathol Appl Neurobiol 30:676–682

Furtado S, Suchowersky O, Rewcastle B, Graham L, Klimek ML, Garber A (1996) Relationship between trinucleotide repeats and neuropathological changes in Huntington's disease. Ann Neurol 39:132–136

Fuster JM (1989) The prefrontal cortex. Raven Press, New York

Garey LJ (1990) Visual system. In: Paxinos G (ed) The human nervous system. Academic Press, San Diego, pp 945–977

Ghez C (1991) The cerebellum. In: Kandel ER, Schwartz JH, Jessell TM (eds) Principles of neural science, 3rd edn. Elsevier, New York, pp 596–607

Giguere M, Goldman-Rakic PS (1988) Mediodorsal nucleus: areal, laminar, and tangential distribution of afferents and efferents in the frontal lobe of rhesus monkeys. J Comp Neurol 277:195–213

Gil JM, Rego AC (2008) Mechanisms of neurodegeneration in Huntington's disease. Eur J Neurosci 27:2803–2820

Gilman S (1992) Cerebellum and motor dysfunction. In: Asbury AK, McKhann GM, McDonald WI (eds) Diseases of the nervous system: clinical neurobiology, 2nd edn. Saunders, Philadelphia, pp 368–389

Goedert M, Jakes R, Crowther RA, Cohen P, Vanmechelen E, Vandermeeren M, Cras P (1994) Epitope mapping of monoclonal antibodies to the paired helical filaments of Alzheimer's disease: identification of phosphorylation sites in tau protein. Biochem J 301:871–877

Goedert M, Clavaguera F, Tolnay M (2010) The propagation of prion-like protein inclusions in neurodegenerative diseases. Trend Neurosci 33:317–325

Golding CV, Danchaivijtr C, Hodgson TL, Trabizi SJ, Kennart C (2006) Identification of an oculomotor biomarker of preclinical Huntington disease. Neurology 67:485–487

Gómez-Tortosa E, Macdonald ME, Friend JC, Taylor SAM, Weiler LJ, Cupples LA, Srinidhi J, Gusella JF, Bird ED, Vonsattel JP, Myers RH (2001) Quantitative neuropathological changes in presymptomatic Huntington's disease. Ann Neurol 49:29–34

Gourfinkel-An I, Cancel G, Duyckaerts C, Faucheux B, Hauw JJ, Trottier Y, Brice A, Agid Y, Hirsch EC (1998) Neuronal distribution of intranuclear inclusions in Huntington's disease with adult onset. Neuroreport 9:1823–1826

Graeber MB, Moran LB (2002) Mechanisms of cell death in neurodegenerative diseases: fashion, fiction, and facts. Brain Pathol 12:385–390

Grimbergen YA, Knol MJ, Bloem BR, Kremer BP, Roos RA, Munneke M (2008) Falls and gait disturbances in Huntington's disease. Mov Disord 23:970–976

Grinberg LT, Heinsen H (2009) Argyrophilic grain disease. An update on a frequent cause of dementia. Dement Neuropsych 3:2–7

Groenewegen HJ (2003) The basal ganglia and motor control. Neural Plast 10:107–120

Gunawardena S, Goldstein LSB (2005) Polyglutamine diseases and transport problems. Arch Neurol 62:46–51

Gutekunst CA, Li SH, Yi H, Mulroy JS, Kuemmerle S, Jones R, Rye D, Ferrante RJ, Hersch SM, Li XJ (1999) Nuclear and neuropil aggregates in Huntington's disease: relationship to neuropathology. J Neurosci 19:2522–2534

Haber SN, Calzavara R (2009) The cortico-basal ganglia integrative network: the role of the thalamus. Brain Res Bull 78:69–74

Hadzi TC, Hendricks AE, Latourelle JC, Lunetta KL, Cupples LA, Gillis T, Mysore JS, Gusella JF, MacDonald ME, Myers RH, Vonsattel JP (2012) Assessment of cortical and striatal involvement in 523 Huntington disease brains. Neurology 79:1708–1715

Halliday G (2004) Substantia nigra and locus coeruleus. In: Paxinos G, Mai JK (eds) The human nervous system, 2nd edn. Elsevier, San Diego, pp 449–463

Halliday GM, McRitchie DA, Macdonald V, Double KL, Trent RJ, Mccusker E (1998) Regional specificity of brain atrophy in Huntington's disease. Exp Neurol 154:663–672

Harper PS (1992) The epidemiology of Huntington's disease. Hum Genet 89:365–376

Harting JK, Huerta MF, Frankfurter AJ, Strominger NL, Royce GJ (1980) Ascending pathways from the monkey superior colliculus: an autoradiographic analysis. J Comp Neurol 192:853–882

Hassel B, Tessler S, Faull RL, Emson PC (2008) Glutamate uptake is reduced in prefrontal cortex in Huntington's disease. Neurochem Res 33:232–237

Hassler R (1982a) Architectonic organization of the thalamic nuclei. In: Schaltenbrand G, Bailey P (eds) Stereotaxy of the human brain. Thieme, Stuttgart/New York, pp 140–180

Hassler R (1982b) Stereotaxic surgery for psychiatric disorders. In: Schaltenbrand G, Bailey P (eds) Stereotaxy of the human brain. Thieme, Stuttgart/New York, pp 570–590

Heathfield KW (1973) Huntington's chorea: a centenary review. Postgrad Med J 49:32–45

Hedreen JC, Peyser CE, Folstein SE, Ross CA (1991) Neuronal loss in layers V and VI of cerebral cortex in Huntington's disease. Neurosci Lett 133:257–261
Heinsen H, Heinsen YL (1991) Serial thick, frozen, gallocyanin stained sections of human central nervous system. J Histotechnol 14:167–173
Heinsen H, Rüb U (1997) Quantitative investigations of neural circuits in Huntington's disease. J Neural Transm 104:1139–1140
Heinsen H, Bauer M, Ulmar G, Gangnus D, Jungkunz G (1992) The entorhinal region in Huntington's disease: a cytoarchitectonic and quantitative investigation in five cases. Clin Neuropathol 11:226
Heinsen H, Strik M, Bauer M, Luther K, Ulmar G, Gangnus D, Jungkunz G, Eisenmenger W, Götz M (1994) Cortical and striatal neurone number in Huntington's disease. Acta Neuropathol 88:320–333
Heinsen H, Rüb U, Gangnus D, Jungkunz G, Bauer M, Ulmar G, Bethke B, Schüler M, Böcker F, Eisenmenger W, Götz M, Strik M (1996) Nerve cell loss in the thalamic centromedian-parafascicular complex in patients with Huntington's disease. Acta Neuropathol 91:161–168
Heinsen H, Rüb U, Bauer M, Ulmar G, Bethke B, Schüler M, Böcker F, Eisenmenger W, Götz M, Korr H, Schmitz C (1999) Nerve cell loss in the thalamic mediodorsal nucleus in Huntington's disease. Acta Neuropathol 97:613–622
Hennerici M, Homberg V, Lange HW (1985) Evoked potentials in patients with Huntington's disease and their offspring. II. Visual evoked potentials. Electroencephalogr Clin Neurophysiol 62:167–176
Herrero MT, Barcia C, Navarro JM (2002) Functional anatomy of thalamus and basal ganglia. Childs Nerv Syst 18:386–404
Hirai T, Jones EG (1989) A new parcellation of the human thalamus on the basis of histochemical staining. Brain Res Rev 14:1–34
Hoche F, Seidel K, Brunt ER, Auburger G, Schöls L, Bürk K, de Vos RA, den Dunnen W, Bechmann I, Egensperger R, Van Broeckhoven C, Gierga K, Deller T, Rüb U (2008) Involvement of the auditory brainstem system in spinocerebellar ataxia type 2 (SCA2), type 3 (SCA3) and type 7 (SCA7). Neuropathol Appl Neurobiol 34:479–491
Hodges A, Strand AD, Aragaki AK, Kuhn A, Sengstag T, Hughes G, Elliston LA, Hartog C, Goldstein DR, Thu D, Hollingsworth ZR, Collin F, Synek B, Holmans PA, Young AB, Wexler N, Delorenzi M, Kooperberg C, Augood SJ, Faull RLM, Olson JM, Jones L, Luthi-Carter R (2006) Regional and cellular gene expression changes in human Huntington's disease brain. Hum Mol Genet 15:965–977
Horn AK (2006) The reticular formation. Prog Brain Res 151:127–155
Horn AKE, Büttner-Ennever JA (1998) Premotor neurons for vertical eye movements in the rostral mesencephalon of monkey and human: Histologic identification by parvalbumin immunostaining. J Comp Neurol 392:413–427
Horn AKE, Büttner-Ennever JA, Wahle P, Reichenberger I (1994) Neurotransmitter profile of saccadic omnipause neurons in nucleus raphe interpositus. J Neurosci 14:2032–2046
Horn AKE, Büttner-Ennever JA, Suzuki A, Henn V (1995) Histological identification of premotor neurons for horizontal saccades in monkeys and man by parvalbumin immunostaining. J Comp Neurol 359:350–563
Horn AKE, Büttner U, Büttner-Ennever JA (1999) Brainstem and cerebellar structures for eye movement generation. Adv Otorhinolaryngol 55:1–25
Huerta MF, Krubitzer LA, Kaas JH (1986) Frontal eye field as defined by intracortical microstimulation in squirrel monkeys, owl monkeys, and macaque monkeys: I. Subcortical connections. J Comp Neurol 253:415–439
Hunt JR (1917) Progressive atrophy of the globus pallidus. Brain 40:58–148
Huntington G (1872) On chorea. Med Surg Reporter 26:317–321
Imarisio S, Carmichael J, Korolchuk V, Chen CW, Saiki S, Rose C, Krishna G, Davies JE, Ttofi E, Underwood BR, Rubinsztein DC (2008) Huntington's disease: from pathology and genetics to potential therapies. Biochem J 412:191–209

Itzhaki RF, Wozniak MA, Appelt DM, Balin BJ (2004) Infiltration of the brain by pathogens causes Alzheimer's disease. Neurobiol Aging 25:619–627

Jelgersma G (1908) Über anatomische Befunde bei Paralyis agitans und bei chronischer Chorea. Neurol Centralblatt 27:995–996

Jellinger KA (2006) Alzheimer 100 – Highlights in the history o Alzheimer research. J Neural Transm 113:1603–1623

Jeste DV, Barban L, Parisi J (1984) Reduced Purkinje cell density in Huntington's disease. Exp Neurol 85:78–86

Jones EG (1985) The thalamus. Plenum Press, New York/London

Jones EG (2003) Two minds. Nature 421:19–20

Josiassen RC, Shagass C, Mancall EL, Roemer RA (1984) Auditory and visual evoked potentials in Huntington's disease. Electroencephalogr Clin Neurophysiol 57:113–118

Jucker M, Walker LC (2011) Pathogenic protein seeding in Alzheimer disease and other neurodegenerative disorders. Ann Neurol 70:532–540

Kato S, Hirano A, Umahara T, Llena JF, Herz F, Ohama E (1992) Ultrastructural and immunohistochemical studies on ballooned cortical neurons in Creutzfeldt-Jakob disease: expression of αB-crystallin, ubiquitin and stress-response protein 27. Acta Neuropathol 84:443–448

Kennard C, Lueck CJ (1989) Oculomotor abnormalities in diseases of the basal ganglia. Rev Neurol 145:587–595

Kiesselbach G (1914) Anatomischer Befund eines Falles von Huntingtonscher Chorea. Monatsschr Psychiat Neurol 35:525–543

Kirkwood SC, Siemers E, Hodes ME, Conneally PM, Christian JC, Foroud T (2000) Subtle changes among presymptomatic carriers of the Huntington's disease gene. J Neurol Neurosurg Psychiatry 69:773–779

Knikou M (2008) The H-reflex as a probe: pathways and pitfalls. J Neurosci Methods 171:1–12

Koeppen AH (1989) The nucleus pontis centralis caudalis in Huntington's disease. J Neurol Sci 91:129–141

Koller WC, Trimble J (1985) The gait abnormality of Huntington's disease. Neurology 35:1450–1454

Kremer B, Weber B, Hayden MR (1992) New insights into the clinical features, pathogenesis and molecular genetics of Huntington disease. Brain Pathol 2:321–335

Kuemmerle S, Gutekunst CA, Klein AM, Li XJ, Li SH, Beal MF, Hersch SM, Ferrante RJ (1999) Huntington aggregates may not predict neuronal death in Huntington's disease. Ann Neurol 46:842–849

Künzle H, Akert K (1977) Efferent connections of cortical area 8 (frontal eye field) in Macaca fascicularis. A reinvestigation using the autoradiographic technique. J Comp Neurol 173:147–164

Kuusisto E, Salminen A, Alafuzoff I (2001) Ubiquitin-binding protein p62 is present in neuronal and glial inclusions in human tauopathies and synucleinopathies. Neuroreport 12:2085–2090

Kuusisto E, Kauppinen T, Alafuzoff I (2008) Use of p62/SQSTM1 antibodies for neuropathological diagnosis. Neuropathol Appl Neurobiol 34:169–180

Labbadia J, Morimoto RI (2013) Huntington's disease: underlying molecular mechanisms and emerging concepts. Trends Biochem Sci 38:378–385

Lace GL, Wharton SB, Ince PG (2007) A brief history of tau: the evolving view of the microtubule-associated protein tau in neurodegenerative diseases. Clin Neuropathol 26:43–58

Landwehrmeyer GB, McNeil SM, Dure LS, Ge P, Aizawa H, Huang Q, Ambrose CM, Duyao MP, Bird ED, Bonilla E (1995) Huntington's disease gene: regional and cellular expression in brain of normal and affected individuals. Ann Neurol 37:218–223

Lange HW (1981) Quantitative changes of telencephalon, diencephalon, and mesencephalon in Huntington's chorea, postencephalitic, and idiopathic Parkinsonism. Verh Anat Ges 73:923–925

Lange HW, Aulich A (1986) Die Hirnatrophie bei der Huntingtonschen Krankheit. In: Oepen H (ed) Die Huntingtonsche Krankheit. Hippokrates, Stuttgart, pp 25–41

Lange H, Thorner G, Hopf A, Schröder KF (1976) Morphometric studies of the neuropathological changes in choreatic diseases. J Neurol Sci 28:401–425
Lanska DJ (2000) George Huntington (1850–1916) and hereditary chorea. J Hist Neurosci 9:76–89
Lanska DJ (2010) The history of movement disorders. In: Finger S, Boller F, Tyler KL (eds) Handbook of clinical neurology, vol 95. Elsevier, Amsterdam, pp 501–546
Lasker AG, Zee DS (1997) Ocular motor abnormalities in Huntington's disease. Vision Res 37:3639–3645
Lasker AG, Zee DS, Hain TC, Folstein SE, Singer HS (1987) Saccades in Huntington's disease: initiation defects and distractibility. Neurology 37:364–370
Lasker AG, Zee DS, Hain TC, Folstein SE, Singer HS (1988) Saccades in Huntington's disease: slowing and dysmetria. Neurology 38:427–431
Lastres-Becker I, Rüb U, Auburger G (2008) Spinocerebellar ataxia 2 (SCA2). Cerebellum 7:115–124
Leak RK (2014) Heat shock proteins in neurodegenerative disorders and aging. J Cell Commun Signal 8:293–310
Leigh RJ, Kennard C (2004) Using saccades as a research tool in the clinical neurosciences. Brain 127:460–477
Leigh RJ, Zee DS (2006) The neurology of eye movements, 4th edn. Oxford University Press, Oxford
Leigh RJ, Newman SA, Folstein SE, Lasker AG, Jensen BA (1983) Abnormal ocular motor control in Huntington's disease. Neurology 33:1268–1275
Leigh RJ, Parhad IM, Clark AW, Buettner-Ennever JA, Folstein SE (1985) Brainstem findings in Huntington's disease. Possible mechanisms for slow vertical saccades. J Neurol Sci 71:247–256
Leopold HC, Doerr M, Oepen G, Thoden U (1982) The effect of cervical and vestibular reflexes on eye movements in Huntington's chorea. Arch Psychiatr Nervenkr 231:227–234
Lewy FH (1923) Die Histopathologie der choreatischen Erkrankungen. Zschr ges Neurol Psychiatr 85:622–658
Li JY, Conforti L (2013) Axonopathy in Huntington's disease. Exp Neurol 246:62–71
Li XJ, Li S (2011) Proteasomal dysfunction in aging and Huntington disease. Neurobiol Dis 43:4–8
Lin YS, Chen CM, Soong BW, Wu YR, Chen HM, Yeh WY, Wu DR, Lin YJ, Poon PW, Cheng ML, Wang CH, Chern Y (2011) Dysregulated brain creatine kinase is associated with hearing impairment in mouse models of Huntington disease. J Clin Invest 121:1519–1523
Lund JC (1860) Chorea Sti Viti i Sætersdalen. Uddrag af Distriktslæge J. C. Lunds Medicinalberetning for 1860. Beretning om Sundhedstilstanden (Norway): 137–138
Lyon IW (1863) Chronic hereditary chorea. Am Med Times 7:289–290
Maat-Schieman ML, Dorsman JC, Smoor MA, Siesling S, van Duinen SG, Verschuuren JJ, den Dunnen JT, van Ommen GJ, Roos RA (1999) Distribution of inclusions in neuronal nuclei and dystrophic neurites in Huntington disease brain. J Neuropathol Exp Neurol 58:129–137
Maat-Schieman M, Roos R, Losekoot M, Dorsman J, Welling-Graafland C, Hegeman-Kleinn I, Broeyer F, Breuning M, van Duinen S (2007) Neuronal intranuclear and neuropil inclusions for pathological assessment of Huntington's disease. Brain Pathol 17:31–37
MacDonald V, Halliday G (2002) Pyramidal cell loss in motor cortices in Huntington's disease. Neurobiol Dis 10:378–386
MacDonald V, Halliday GM, Trent RJ, McCusker EA (1997) Significant loss of pyramidal neurons in the angular gyrus of patients with Huntington's disease. Neuropathol Appl Neurobiol 23:492–495
Mann DM, Oliver R, Snowden JS (1993) The topographic distribution of brain atrophy in Huntington's disease and progressive supranuclear palsy. Acta Neuropathol 85:553–559
Margolis RL, Ross CA (2003) Diagnosis of Huntington disease. Clin Chem 49:1726–1732
Markowitsch HJ (1982) Thalamic mediodorsal nucleus and memory: a critical evaluation of studies in animals and man. Neurosci Biobehav Rev 6:351–380

May PJ (2006) The mammalian superior colliculus: laminar structure and connections. Prog Brain Res 151:321–378

McCaughey WTE (1961) The pathologic spectrum of Huntington's chorea. J Nerv Ment Dis 133:91–103

McFarland NR, Haber SN (2002) Thalamic relay nuclei of the basal ganglia form both reciprocal and nonreciprocal cortical connections, linking multiple frontal cortical areas. J Neurosci 22:117–132

McLardy T (1948) Projection of the centromedian nucleus of the human thalamus. Brain 71:290–303

McLeod JG (1969) H-reflex studies in patients with cerebellar disorders. J Neurol Neurosurg Psychiatry 32:21–27

Mesulam MM (1998) From sensation to cognition. Brain 121:1013–1052

Meyer A, Beck F, McLardy T (1947) Prefrontal leucotomy: a neuroanatomical report. Brain 70:18–49

Millecamps S, Julien JP (2013) Axonal transport deficits and neurodegenerative diseases. Nat Rev Neurosci 14:161–176

Mink JW (1996) The basal ganglia: Focused selection and inhibition of competing motor programs. Prog Neurobiol 50:381–425

Misiaszek JE (2003) The H-reflex as a tool in neurophysiology: its limitations and uses in understanding nervous system function. Muscle Nerve 28:144–160

Modregger J, DiProspero NA, Charles V, Tagle DA, Plomann M (2002) PACSIN 1 interacts with huntingtin and is absent from synaptic varicosities in presymptomatic Huntington's disease brains. Hum Mol Genet 11:2547–2558

Montoya A, Price BH, Menear M, Lepage M (2006) Brain imaging and cognitive dysfunctions in Huntington's disease. J Psychiatry Neurosci 31:21–29

Moore JK, Linthicum FH (2004) Auditory system. In: Paxinos G, Maik JK (eds) The human nervous system, 2nd edn. Elsevier, San Diego, pp 1241–1279

Morecraft RJ, Geula C, Mesulam MM (1992) Cytoarchitecture and neural afferents of orbitofrontal cortex in the brain of the monkey. J Comp Neurol 323:341–358

Morel A, Magnin M, Jeanmonod D (1997) Multiarchitectonic and stereotactic atlas of the human thalamus. J Comp Neurol 387:588–630

Myers RH (2004) Huntington's disease genetics. NeuroRx 1:255–262

Myers RH, Vonsattel JP, Stevens TJ, Cupples LA, Richardson EP, Martin JB, Bird ED (1988) Clinical and neuropathologic assessment of severity in Huntington's disease. Neurology 38:341–347

Myers RH, Vonsattel JP, Paskevich PA, Kiely DK, Stevens TJ, Cupples LA, Richardson EPJ, Bird ED (1991) Decreased neuronal and increased oligodendroglial densities in Huntington's disease caudate nucleus. J Neuropathol Exp Neurol 50:729–742

Nakano K, Kayahara T, Tsutsumi T, Ushiro H (2000) Neural circuits and functional organization of the striatum. J Neurol 247(Suppl 5):V1–V5

Neustaedter M (1933) Concerning the striatal localization in chronic progressive chorea. With a report of three cases, two of the Huntington type in siblings and one senile arteriosclerotic, with necropsies. J Nerv Ment Dis 78:470–491

Nopoulos PC, Aylward EH, Ross CA, Johnson HJ, Magnotta VA, Juhl AR, Pierson RK, Mills J, Langbehn DR, Paulsen JS (2010) Cerebral cortex structure in prodromal Huntington disease. Neurobiol Dis 40:544–554

Norrby E (2011) Prions and protein-folding diseases. J Intern Med 270:1–14

Oepen G, Hillesheimer W (1981) Qualitative H-reflex testing in Huntington's disease. Arch Psychiatr Nervenkr 230:221–226

Oepen G, Ostertag C (1981) Diagnostic value of CT in patients with Huntington's chorea and their offspring. J Neurol 225:189–196

Oepen G, Clarenbach P, Thoden U (1981) Disturbance of eye movements in Huntington's chorea. Arch Psychiatr Nervenkr 229:205–213

Oepen G, Mohr U, Willmes K, Thoden U (1985) Huntington's disease: visuomotor disturbance in patients and offspring. J Neurol Neurosurg Psychiatry 48:426–433
Oliveira JMA (2010) Nature and cause of mitochondrial dysfunction in Huntington's disease: focusing on huntingtin and the striatum. J Neurochem 114:1–12
Oorschot DE (1994) Are you using neuronal densities, synaptic densities or neurochemical densities as your definitive data? There is a better way to go. Prog Neurobiol 44:233–247
Ortega Z, Díaz-Hernández M, Lucas JJ (2007) Is the ubiquitin-proteasome system impaired in Huntington's disease? Cell Mol Life Sci 64:2245–2257
Oyanagi K, Takeda S, Takahashi H, Ohama E, Ikuta F (1989) A quantitative investigation of the substantia nigra in Huntington's disease. Ann Neurol 26:13–19
Pandey M, Mohanakumar KP, Usha R (2010) Mitochondrial functional alterations in relation to pathophysiology of Huntington's disease. J Bioenerg Biomembr 42:217–226
Parent A (1990) Extrinsic connections of the basal ganglia. Trends Neurosci 13:254–258
Parent A, Hazrati LN (1995) Functional anatomy of the basal ganglia. 1. The cortico-basal ganglia-thalamo-cortical loop. Brain Res Rev 20:91–127
Pattison LR, Kotter MR, Fraga D, Bonelli RM (2006) Apoptotic cascades as possible targets for inhibiting cell death in Huntington's disease. J Neurol 253:1137–1142
Paul S (2008) Dysfunction of the ubiquitin-proteasome system in multiple disease conditions: therapeutic approaches. Bioessays 3:1172–1184
Paulus W, Jellinger K (1991) The neuropathologic basis of different clinical subgroups of Parkinson's disease. J Neuropathol Exp Neurol 50:743–755
Peltsch A, Hoffman A, Armstrong I, Pari G, Munoz DP (2008) Saccadic impairments in Huntington's disease. Exp Brain Res 186:457–469
Penney JB Jr, Vonsattel JP, MacDonald ME, Gusella JF, Myers RH (1997) CAG repeat number governs the development rate of pathology in Huntington's disease. Ann Neurol 41:689–692
Petersen A, Gil J, Maat-Schieman ML, Bjorkqvist M, Tanila H, Araujo IM, Smith R, Popovic N, Wierup N, Norlen P, Li JY, Ross RA, Sundler F, Mulder H, Brundin P (2005) Orexin loss in Huntington's disease. Hum Mol Genet 14:39–47
Pfeiffer JAF (1913) A contribution to the pathology of chronic progressive chorea. Brain 35:276–292
Pillon B, Deweer B, Agid Y, Dubois B (1993) Explicit memory in Alzheimer's, Huntington's, and Parkinson's diseases. Arch Neurol 50:374–379
Rapoport SI (1990) Integrated phylogeny of the primate brain, with special reference to humans and their diseases. Brain Res Rev 15:267–294
Renner M, Melki R (2014) Protein aggregation and prionopathies. Pathol Biol 62:162–168
Richardson EP Jr (1990) Third Dorothy S. Russell memorial lecture. Huntington's disease: some recent neuropathological studies. Neuropathol Appl Neurobiol 16:451–460
Riess O, Rüb U, Pastore A, Bauer P, Schöls L (2008) SCA3: neurological features, pathogenesis and animal models. Cerebellum 7:125–137
Robinson FR, Fuchs AF (2001) The role of the cerebellum in voluntary eye movements. Annu Rev Neurosci 24:981–1004
Rodda RA (1981) Cerebellar atrophy in Huntington's disease. J Neurol Sci 50:147–157
Roos RA, Pruyt JF, de Vries J, Bots GT (1985) Neuronal distribution in the putamen in Huntington's disease. J Neurol Neurosurg Psychiatry 48:422–425
Rosas HD, Liu AK, Hersch S, Glessner M, Ferrante RJ, Salat DH, van der Kouwe A, Jenkins BG, Dale AM, Fischl B (2002) Regional and progressive thinning of the cortical ribbon in Huntington's disease. Neurology 58:695–701
Rosas HD, Koroshetz WJ, Chen YI, Skeuse C, Vangel M, Cudkowicz ME, Kaplan K (2003) Evidence for more widespread cerebral pathology in early HD: an MRI-based morphometric analysis. Neurology 60:1615–1620
Ross CA, Tabrizi SJ (2011) Huntington's disease: from molecular pathogenesis to clinical treatment. Lancet Neurol 10:83–98
Rüb U, Del Tredici K, Schultz C, Büttner-Ennever JA, Braak H (2001) The premotor region essential for rapid vertical eye movements shows early involvement in Alzheimer's disease-related cytoskeletal pathology. Vision Res 41:2149–2156

Rüb U, Brunt ER, Gierga K, Schultz C, Paulson H, de Vos RA, Braak H (2003a) The nucleus raphe interpositus in spinocerebellar ataxia type 3 (Machado-Joseph disease). J Chem Neuroanat 25:115–127

Rüb U et al (2003b) Thalamic involvement in a spinocerebellar ataxia type 2 (SCA2) and a spinocerebellar ataxia type 3 (SCA3) patient, and its clinical relevance. Brain 126:2257–2272

Rüb U, Brunt ER, de Vos RA, Del Turco D, Del Tredici K, Gierga K, Schultz C, Ghebremedhin E, Bürk K, Auburger G, Braak H (2004a) Degeneration of the central vestibular system in spinocerebellar ataxia type 3 (SCA3) patients and its possible clinical significance. Neuropathol Appl Neurobiol 30:402–414

Rüb U, Bürk K, Schöls L, Brunt ER, de Vos RA, Diaz GO, Gierga K, Ghebremedhin E, Schultz C, Del Turco D, Mittelbronn M, Auburger G, Deller T, Braak H (2004b) Damage to the reticulotegmental nucleus of the pons in spinocerebellar ataxia type 1, 2, and 3. Neurology 63:1258–1263

Rüb U, Gierga K, Brunt ER, de Vos RA, Bauer M, Schöls L, Bürk K, Auburger G, Bohl J, Schultz C, Vuksic M, Burbach GJ, Braak H, Deller T (2005) Spinocerebellar ataxias types 2 and 3: degeneration of the precerebellar nuclei isolates the three phylogenetically defined regions of the cerebellum. J Neural Transm 112:1523–1545

Rüb U, Brunt ER, Deller T (2008a) New insights into the pathoanatomy of spinocerebellar ataxia type 3 (Machado-Joseph disease). Curr Opin Neurol 21:111–116

Rüb U, Jen JC, Braak H, Deller T (2008b) Functional neuroanatomy of the human premotor oculomotor brainstem nuclei: insights from postmortem and advanced in vivo imaging studies. Exp Brain Res 187:167–180

Rüb U, Heinsen H, Brunt ER, Landwehrmeyer B, den Dunnen WF, Gierga K, Deller T (2009) The human premotor oculomotor brainstem system – can it help to understand oculomotor symptoms in Huntington's disease? Neuropathol Appl Neurobiol 35:4–15

Rüb U, Hoche F, Brunt ER, Heinsen H, Seidel K, Del Turco D, Paulson HL, Bohl J, von Gall C, Vonsattel JP, Korf HW, den Dunnen WF (2013a) Degeneration of the cerebellum in Huntington's disease (HD): possible relevance for the clinical picture and potential gateway to pathological mechanisms of the disease process. Brain Pathol 23:165–177

Rüb U, Schöls L, Paulson H, Auburger G, Kermer P, Jen JC, Seidel K, Korf HW, Deller T (2013b) Clinical features, neurogenetics and neuropathology of the polyglutamine spinocerebellar ataxias type 1, 2, 3, 6 and 7. Prog Neurobiol 104:38–66

Rüb U, Hentschel M, Stratmann K, Brunt E, Heinsen H, Seidel K, Bouzrou M, Auburger G, Paulson H, Vonsattel JP, Lange H, Korf HW, den Dunnen W (2014a) Huntington's disease (HD): degeneration of select nuclei, widespread occurrence of neuronal nuclear and axonal inclusions in the brainstem. Brain Pathol 24:247–260

Rüb U, Seidel K, Vonsattel JP, Lange HW, Eisenmenger W, Götz, M, Del Turco, D, Bouzrou M, Korf HW, Heinsen H (2014b) Huntington's disease (HD): neurodegeneration of Brodmann's primary visual area 17 (BA17). Brain Pathol [Epub ahead of print]

Sadikot AF, Rymar VV (2009) The primate centromedian-parafascicular complex: anatomical organization with a note on neuromodulation. Brain Res Bull 78:122–130

Sadikot AF, Parent A, Franqois C (1992a) The centre median and parafascicular thalamic nuclei project respectively to the sensorimotor and associative-limbic striatal territories in the squirrel monkey. Brain Res 510:161–165

Sadikot AF, Parent A, Francois C (1992b) Efferent connections of the centromedian and parafascicular thalamic nuclei in the squirrel monkey: a PHA-L study of subcortical projections. J Comp Neurol 315:137–159

Salmon DP, Kwo on Yuen, PF, Heindel WC, Butters N, Thal LJ (1989) Differentiation of Alzheimer's disease and Huntington's disease with the Dementia Rating Scale. Arch Neurol 46:1204–1208

Sapp E, Penney J, Young A, Aronin N, Vonsattel JP, Difiglia M (1999) Axonal transport of N-terminal huntingtin suggests early pathology of corticostriatal projections in Huntington disease. J Neuropathol Exp Neurol 58:165–173

Schapira AH, Olanow CW, Greenamyre JT, Bezard E (2014) Slowing of neurodegeneration in Parkinson's disease and Huntington's disease: future therapeutic perspectives. Lancet 384:545–555

Scherzed W, Brunt ER, Heinsen H, de Vos RA, Seidel K, Bürk K, Schöls L, Auburger G, Del Turco D, Deller T, Korf HW, den Dunnen WF, Rüb U (2012) Pathoanatomy of cerebellar degeneration in spinocerebellar ataxia type 2 (SCA2) and type 3 (SCA3). Cerebellum 11:749–760

Schmitz C, Hof PR (2005) Design-based stereology in neuroscience. Neuroscience 130:813–831

Schroeder K (1931) Zur Klinik und Pathologie der Huntingtonschen Krankheit. J Psychol Neurol 43:183–201

Schulte J, Littleton JT (2011) The biological function of the Huntingtin protein and its relevance to Huntington's Disease pathology. Curr Trends Neurol 5:65–78

Seibenhener ML, Babu JR, Geetha T, Wong HC, Krishna NR, Wooten MW (2004) Sequestosome 1/p62 is a polyubiquitin chain binding protein involved in ubiquitin proteasome degradation. Mol Cell Biol 24:8055–8068

Seidel K, Brunt ER, de Vos RA, Dijk F, van der Want HJ, Rüb U, Den Dunnen WF (2009) The p62 antibody reveals various cytoplasmic protein aggregates in spinocerebellar ataxia type 6. Clin Neuropathol 28:344–349

Seidel K, Den Dunnen WF, Schultz C, Paulson H, Frank S, de Vos RA, Brunt ER, Deller T, Kampinga HH, Rüb U (2010) Axonal inclusions in spinocerebellar ataxia type 3. Acta Neuropathol 120:449–460

Seidel K, Siswanto S, Brunt ER, den Dunnen W, Korf HW, Rüb U (2012) Brain pathology of spinocerebellar ataxias. Acta Neuropathol 124:1–21

Seidel K, Mahlke J, Siswanto S, Krüger R, Heinsen H, Auburger G, Bouzrou M, Grinberg LT, Wicht H, Korf HW, den Dunnen W, Rüb U (2015) The brainstem pathologies of Parkinson's disease and dementia with lewy bodies. Brain Pathol 25:121–135

Selemon LD, Rajkowska G, Goldman-Rakic PS (2004) Evidence for progression in frontal cortical pathology in late-stage Huntington's disease. J Comp Neurol 468:190–204

Shaw G, Osborn M, Weber K (1981) An immunofluorescence microscopical study of the neurofilament triplet proteins, vimentin and glial fibrillary acidic protein within the adult rat brain. Eur J Cell Biol 26:68–82

Sheperd GM (2013) Corticostriatal connectivity and its role in disease. Nat Rev Neurosci 14:278–291

Shoulson I, Young AB (2011) Milestones in huntington disease. Mov Disord 26:1127–1133

Sieradzan KA, Mann DM (2001) The selective vulnerability of nerve cells in Huntington's disease. Neuropathol Appl Neurobiol 27:1–21

Sieradzan KA, Mechan AO, Jones L, Wanker EE, Nukina N, Mann DM (1999) Huntington's disease intranuclear inclusions contain truncated, ubiquitinated huntingtin protein. Exp Neurol 156:92–99

Smith R, Klein P, Koc-Schmitz Y, Waldvogel HJ, Faull RL, Brundin P, Plomann M, Li JY (2007) Loss of SNAP-25 and rabphilin 3a in sensory-motor cortex in Huntington's disease. J Neurochem 103:115–123

Sommer MA, Wurtz RH (2004) What the brain stem tells the frontal cortex. I. Oculomotor signals sent from superior colliculus to frontal eye field via mediodorsal thalamus. J Neurophysiol 91:1381–13402

Sotrel A, Paskevich PA, Kiely DK, Bird ED, Williams RS, Myers RH (1991) Morphometric analysis of the prefrontal cortex in Huntington's disease. Neurology 41:1117–1121

Sotrel A, Williams RS, Kaufmann WE, Myers RH (1993) Evidence for neuronal degeneration and dendritic plasticity in cortical pyramidal neurons of Huntington's disease: a quantitative Golgi study. Neurology 43:2088–2096

Spielmeyer W (1926) Die anatomische Krankheitsforschung am Beispiel einer Huntingtonschen Chorea mit Wilsonschem Symptomenbild. Zschr ges Neurol Psychiatr 101:701–728

Stanton GB, Goldberg ME, Bruce CJ (1988) Frontal eye field efferents in the macaque monkey. I Subcortical pathways and topography of striatal and thalamic terminal fields. J Comp Neurol 271:473–492

Starr A (1967) A disorder of rapid eye movements in Huntington's chorea. Brain 90:545–564

Stier E (1903) Zur pathologischen Anatomie der Huntingtonschen Chorea. Arch f Psych 37:62–86

Stover NP, Watts RL (2001) Corticobasal degeneration. Sem Neurol 21:49–58
Tabrizi SJ, Scahill RI, Durr A, Roos RA, Leavitt BR, Jones R, Landwehrmeyer GB, Fox NC, Johnson H, Hicks SL, Kennard C, Craufurd D, Frost C, Langbehn DR, Reilmann R, Stout JC (2011) Biological and clinical changes in premanifest and early stage Huntington's disease in the TRACK-HD study: the 12-month longitudinal analysis. Lancet Neurol 10:31–42
Tanner CM, Goldman SM (1994) Epidemiology of movement disorders. Curr Opin Neurol 7:340–345
Terplan K (1924) Zur pathologischen Anatomie der chonischen progressiven Chorea. Arch Path Anat 252:146–176
The Huntington's Disease Collaborative Research Group (1993) A novel gene containing a trinucleotide repeat that is expanded and unstable on Huntington's disease chromosomes. Cell 72:971–983
Thu DC, Oorschot DE, Tippett LJ, Nana AL, Hogg VM, Synek BJ, Luthi-Carter R, Waldvogel HJ, Faull RL (2010) Cell loss in the motor and cingulate cortex correlates with symptomatology in Huntington's disease. Brain 133:1094–1110
Tian JR, Zee DS, Lasker AG, Folstein SE (1991) Saccades in Huntington's disease: predictive tracking and interaction between release of fixation and initiation of saccades. Neurology 41:875–881
Trottier Y, Lutz Y, Stevanin G, Imbert DD, Cancel G, Saudou F, Weber C, David G, Tora L, Agid Y, Brice A, Mandel JL (1995) Polyglutamine expansion as a pathological epitope in Huntington's disease and four dominant cerebellar ataxias. Nature 378:403–406
U.S.-Venezuela Collaborative Research Project, Wexler NS (2004) Venezuelan kindreds reveal that genetic and environmental factors modulate Huntington's disease age of onset. Proc Natl Acad Sci U S A 101:3498–3503
Valera AG, Diaz-Hernandez M, Hernandez F, Ortega Z, Lucas JJ (2005) The ubiquitin-proteasome system in Huntington's disease. Neuroscientist 11:583–594
van Wamelen DJ, Aziz NA, Roos RA, Swaab DF (2014) Hypothalamic alterations in Huntington's disease patients: comparison with genetic rodent models. J Neuroendocrinol 26:761–775
Vogt C, Vogt O (1920) Zur Lehre der Erkrankungen des striären Systems. J Psychol Neurol 25(Suppl 3):631–846
Vogt C, Vogt O (1942) Morphologische Gestaltungen unter normalen und pathogenen Bedingungen. Ein hirnanatomischer Beitrag zu ihrer Kenntnis. J Psychol Neurol 50:161–524
von Economo C, Koskinas GN (1925) Die Cytoarchitekonik der Hirnrinde des erwachsenen Menschen. Julius Springer, Wien Berlin
Vonsattel JP (2008) Huntington disease models and human neuropathology: similarities and differences. Acta Neuropathol 115:55–69
Vonsattel JP, DiFiglia M (1998) Huntington disease. J Neuropathol Exp Neurol 57:369–384
Vonsattel JP, Myers RH, Stevens TJ, Ferrante RJ, Bird ED, Richardson EP Jr (1985) Neuropathological classification of Huntington's disease. J Neuropathol Exp Neurol 44:559–577
Voogd J (2003) The human cerebellum. J Chem Neuroanat 26:243–252
Voogd J, Barmack NH (2006) Oculomotor cerebellum. Prog Brain Res 151:231–268
Voogd J, Feirabend HKP, Schoen JHR (1990) Cerebellum and precerebellar nuclei. In: Paxinos G (ed) The human central nervous system. Academic Press, San Diego, pp 321–386
Voogd J, Schraa-Tam CKL, van der Geest JN, De Zeeuw CI (2010) Visuomotor cerebellum in human and nonhuman primates. Cerebellum 11:392–410
Walker AE (1982) Normal and pathological physiology of the thalamus. In: Schaltenbrand G, Walker AE (eds) Stereotaxy of the human brain. Anatomical, physiological and clinical applications. Thieme, Stuttgart/New York, pp 181–217
Walker FO (2007a) Huntington's disease. Semin Neurol 27:143–150
Walker FO (2007b) Huntington's disease. Lancet 369:218–228
Waters CO (1842) Letter on chorea. In: Dunglison R (ed) The practice of medicine, vol 2. Lea and Blanchard, Philadelphia, pp 312–313
West MJ, Gundersen HJ (1990) Unbiased stereological estimation of the number of neurons in the human hippocampus. J Comp Neurol 296:1–22

Whitney ER, Kemper TL, Rosene DL, Bauman ML, Blatt GJ (2008) Calbindin-D28k is a more reliable marker of human Purkinje cells than standard Nissl stains: a stereological experiment. J Neurosci Methods 168:42–47

Winograd-Gurvich CT, Georgiou-Karistianis N, Evans A, Millist L, Bradshaw JL, Churchyard A, Chiu E, White OB (2003) Hypometric primary saccades and increased variability in visually-guided saccades in Huntington's disease. Neuropsychologia 41:1683–1692

Wooten MW, Hu X, Babu JR, Seibenhener ML, Geetha T, Paine MG, Wooten MC (2006) Signaling, polyubiquitination, trafficking, and inclusions: sequestosome 1/p62's role in neurodegenerative disease. J Biomed Biotechnol 2006:1–12. Article ID 62079

Yanagisawa N (1992) The spectrum of motor disorders in Huntington's disease. Clin Neurol Neurosurg 94(Suppl):182–184

Zilles K (1990) Cortex. In: Paxinos G (ed) The human nervous system. Academic Press, San Diego, pp 757–780

MIX
Papier aus verantwortungsvollen Quellen
Paper from responsible sources
FSC® C105338

FSC
www.fsc.org

If you have any concerns about our products,
you can contact us on
ProductSafety@springernature.com

In case Publisher is established outside the EU,
the EU authorized representative is:
Springer Nature Customer Service Center GmbH
Europaplatz 3, 69115 Heidelberg, Germany

Printed by Libri Plureos GmbH
in Hamburg, Germany